Lecture Notes in Mathematics

Edited by A. Dold and B. Eckmann

Subseries: Mathematisches Institut der Universität und Max-Planck-Institut für Mathematik, Bonn – vol. 12

Adviser: F. Hirzebruch

1293

Wolfgang Ebeling

The Monodromy Groups of Isolated Singularities of Complete Intersections

Springer-Verlag

Berlin Heidelberg New York London Paris Tokyo

Author

Wolfgang Ebeling
Mathematisches Institut der Universität
Wegelerstr. 10, 5300 Bonn 1, Federal Republic of Germany

Mathematics Subject Classification (1980): 14 B 05, 14 D 05, 14 M 10, 32 B 30, 57 R 45; 10 C 30, 10 E 45, 20 H 15, 51 F 15

ISBN 3-540-18686-7 Springer-Verlag Berlin Heidelberg New York
ISBN 0-387-18686-7 Springer-Verlag New York Berlin Heidelberg

© Springer-Verlag Berlin Heidelberg 1987
Printed in Germany

Printing and binding: Druckhaus Beltz, Hemsbach/Bergstr.
2146/3140-543210

For Bettina, Bastian, and Mirja

INTRODUCTION

For the analysis of hypersurface singularities and their deformations the Milnor lattice and the invariants associated with it play an important rôle. The middle homology group of a Milnor fibre of the singularity is a free abelian group, which is endowed with a bilinear form, the intersection form. This bilinear form is symmetric or skew-symmetric, if the dimension is even or odd respectively. This group with this additional structure is called the Milnor lattice H. The monodromy group Γ of the singularity is a subgroup of the automorphism group of this lattice. It is generated by reflections, respectively symplectic transvections, corresponding to certain elements of the Milnor lattice, the vanishing cycles, which are defined in a geometric way. These form the set $\Delta \subset H$ of vanishing cycles. The monodromy group is already generated by the respective automorphisms corresponding to the elements of certain geometrically distinguished bases of vanishing cycles. The set of Dynkin diagrams (or intersection diagrams) corresponding to such bases yields another invariant. A survey of the relations between these invariants and their importance for the deformation theory of the singularities in the hypersurface case is given by E. Brieskorn in his expository article [Brieskorn$_3$].

By stabilizing we can restrict ourselves to the symmetric case in the hypersurface situation. For the simple hypersurface singularities the monodromy groups and the sets of vanishing cycles are finite, and it is well-known that they coincide with the Weyl groups and root systems respectively corresponding to the classical Dynkin diagrams [1)] of type A_m, D_m, E_6, E_7, or E_8. In the general (symmetric) case there arises the question about the nature of these infinite reflection groups and these infinite subsets of the Milnor lattice and about a characterization of these invariants. The work on this question was stimulated by a result of H. Pinkham [Pinkham$_1$] of the year 1977. Pinkham showed that the monodromy groups of the fourteen exceptional unimodal hypersurface singularities are arithmetic. In the sequel we were able to prove that the result of Pinkham is true for large classes of hypersurface

1) The notation "Dynkin diagrams" for these graphs was introduced by N. Bourbaki [Bourbaki$_2$], and is commonly used for these graphs since. Usually one also denotes the more general intersection diagrams of arbitrary hypersurface singularities by this name. We shall also use this notation in this monograph. Unfortunately, this notation does not seem to be historically justified. More accurately, these graphs should be attributed to H.S.M. Coxeter. However, Bourbaki calls the closely related graphs associated with Coxeter systems Coxeter graphs. We follow Bourbaki and reserve this notation for these graphs, too (cf. Remark 4.1.4, Example 5.2.4, and Chapter 5.5).

singularities, using a theorem of M. Kneser. On the other hand it was clear from the beginning that this result could not be true for all hypersurface singularities. One can find exceptions among the other unimodal hypersurface singularities, which appear at the beginning of Arnol'd's classification, namely among the hyperbolic singularities forming the $T_{p,q,r}$-series.

The Milnor lattices, monodromy groups, and vanishing cycles are more generally defined for isolated singularities of complete intersections. Therefore one can also consider the above question in this larger context. The aim of this monograph is to explore these invariants for this general class of singularities, which embraces the hypersurface singularities. This investigation is centred upon the above question.

In order to be able to study these invariants, we first elaborate suitable procedures to compute these invariants. Here we are guided by the hypersurface case. In this case one uses Dynkin diagrams corresponding to geometric bases of vanishing cycles for the calculation of these invariants. However, the notion of a geometric basis of vanishing cycles cannot be readily transferred to the general case. Though one can also consider the corresponding systems of vanishing cycles in the complete intersection case, they are no longer linearly independent, unless we are in the hypersurface case, and one also has to take the linear relations among these cycles into account. We introduce an appropriate notion of a Dynkin diagram for an isolated complete intersection singularity. We present methods to compute such Dynkin diagrams.

We apply these methods to determine Dynkin diagrams and the invariants derived from these graphs for the singularities at the lowest levels of the hierarchy of isolated complete intersection singularities. Using these results, we are in particular able to classify the isolated complete intersection singularities with a definite, parabolic, or hyperbolic intersection form. This extends the corresponding classification of Arnol'd in the hypersurface case [Arnol'd_1] to the case of complete intersections. Moreover, one discovers many interesting relations in this way.

As the main result of this monograph we can answer the question about a characterization of the monodromy groups and of the vanishing cycles in the symmetric case completely. We can show that the monodromy groups and vanishing cycles of all even-dimensional isolated complete intersection singularities except the hyperbolic singularities can be described in a purely arithmetic way, i.e. only in terms of the Milnor lattice. In particular the simple parts of the monodromy groups are arithmetic in these cases. For the hyperbolic singularities one can give another, though non-arithmetic, description of these invariants. These results can also be applied to global monodromy groups. So for example one can prove that the monodromy group of the universal family of projective complete intersections of a fixed (even) dimension and multidegree is arithmetic. The proof of these results is reduced to algebraic results on reflection groups corresponding to vanishing lattices by means of our calculation of Dynkin diagrams.

We now give a survey of the contents of this work in detail. The monograph is divided into five chapters.

In the first chapter we introduce the invariants of isolated singularities of complete intersections to be considered later. Here we alter the point of view compared to the hypersurface case slightly. In the general case we do not regard merely the Milnor lattice, but a short exact sequence of lattices

$$0 \longrightarrow H' \longrightarrow \hat{H} \longrightarrow H \longrightarrow 0,$$

including the Milnor lattice H, as the basic invariant. This sequence is defined as follows (cf. Section 1.1). Let (X,x) be an isolated singularity of an n-dimensional complete intersection, and let $F = (F_1,\dots,F_p) : (\mathbb{C}^{n+p},0) \longrightarrow (\mathbb{C}^p,0)$ be the semi-universal deformation of (X,x). We choose a line ℓ in the base space $\mathbb{C}^p$ through the origin which intersects the discriminant of F at the origin transversally. Without loss of generality we assume that the coordinates of $\mathbb{C}^p$ are chosen in such a way that this line coincides with the last coordinate axis. Then $F' = (F_1,\dots,F_{p-1}):(\mathbb{C}^{n+p},0) \longrightarrow (\mathbb{C}^{p-1},0)$ defines an isolated singularity (X',x). Let X_s and X'_t be the Milnor fibres of (X,x) and (X',x) respectively. Then the above sequence is part of the long exact (reduced) homology sequence of the pair (X'_t,X_s). This means that $H = \tilde{H}_n(X_s,\mathbb{Z})$, $H' = H_{n+1}(X'_t,\mathbb{Z})$, and $\hat{H}$ is the relative homology group $H_{n+1}(X'_t,X_s)$. On these modules we consider the bilinear forms induced by the intersection form on H.

For each invariant associated with the Milnor lattice H we can then consider a corresponding relative invariant which is related to the lattice $\hat{H}$. In this way the vanishing cycles are associated with the thimbles which were already considered by Lefschetz (Section 1.2). These form the set $\hat{\Delta}$. To the monodromy group Γ corresponds the relative monodromy group $\hat{\Gamma}$ (Section 1.3). The natural generalization of the notion of a weakly or strongly distinguished basis of vanishing cycles in the hypersurface case is considered in Section 1.4: This is a weakly or strongly distinguished basis of $\hat{H}$ consisting of thimbles. So the Dynkin diagrams which are introduced in Section 1.5 are Dynkin diagrams corresponding to weakly or strongly distinguished bases of thimbles of $\hat{H}$. As in the hypersurface case they are not uniquely determined, and we also study the possible transformations of these diagrams in this section. At the end of Section 1.5 we show the invariance of the introduced objects. A Dynkin diagram corresponding to a strongly distinguished basis of thimbles determines the remaining relative invariants, in particular also a special element of the relative monodromy group $\hat{\Gamma}$, namely the relative monodromy operator $\hat{c}$. We discuss in Section 1.6 to what extent one can get informations about the module H', and hence about the whole fundamental exact sequence above, from the knowledge of $\hat{H}$ and of the relative monodromy operator $\hat{c}$.

In Chapter 2 we describe our methods for the computation of Dynkin diagrams corresponding to strongly distinguished bases of thimbles. We derive a generalization of a procedure of Gabrielov in the hypersurface case [Gabrielov$_3$] (Section 2.2). Our calculations are essentially based on this method. This procedure allows us to reduce the calculation of Dynkin diagrams to the calculation of Dynkin diagrams for simpler singularities. Here the polar curve of the singularity plays an important rôle. The necessary definitions and facts about polar curves and polar invariants in the case of complete intersections are collected in Section 2.1. The simplest singularities of complete intersections which are not hypersurfaces are the isolated singularities of intersections of two quadrics. For such a singularity H. Hamm has given a basis of the Milnor lattice H [Hamm$_2$]. In Section 2.3 we show that the basis of Hamm consists of vanishing cycles and that these cycles bound the thimbles of a strongly distinguished basis of $\hat{H}$. We compute the Dynkin diagram corresponding to this basis and analyze the invariants of this special singularity by means of suitable transformations of this Dynkin diagram. It turns out that there exists a close relation to K. Saito's theory of extended affine root systems [Saito$_1$], which we discuss in Section 2.4. Finally Section 2.5 deals with another method to compute Dynkin diagrams. This is a generalization of a method of F. Lazzeri in the hypersurface case to determine the intersection matrix using the relations of the fundamental group of the complement of the discriminant. We explain this method by means of an example which is essential for the later applications, but where the application of the procedure of Section 2.2 already leads to considerable difficulties.

In Chapter 3 we apply the method of Chapter 2.2 to calculate Dynkin diagrams for some special singularities. Here we only consider singularities which are given by map-germs $f : (\mathbb{C}^{n+2},0) \rightarrow (\mathbb{C}^2,0)$ with $df(0) = 0$ and with regular 2-jet. This means in particular that H' has rank 1 in this case. In Section 1.1 we consider the classification of these singularities of any dimension, and we introduce the classes of singularities which will be considered later. Part of them will play a rôle in Chapter 4. These classes are characterized by the Segre symbol of the 2-jet. At the beginning of the classification one finds after the intersections of two quadrics the n-dimensional singularities $T^n_{2,q,2,s}$ (Segre symbols $\{1,...,1,2\}$ and $\{1,...,1,(1,1)\}$). In Section 3.2 we explain in detail how one can compute Dynkin diagrams for these singularities using the method of Chapter 2.2. It turns out that these singularities are hyperbolic in the even-dimensional case. Section 3.3 is devoted to two other classes of n-dimenional singularities, namely the singularities of the $J^{(n-1)}$- and $K^{(n-1)}$-series (Segre symbols $\{1,...,1,3\}$ and $\{1,...,1,(1,2)\}$ respectively). Here the calculation of Dynkin diagrams follows the pattern of Section 3.2. The remaining sections of Chapter 3 are not needed for Chapter 4. In Section 3.4 we focus our attention on the case of curves and consider in particular Dynkin diagrams for the simple space curve singularities. The surface case is considered more fully in Sections 3.5 and 3.6. Section 3.6 deals in particular with the triangle singularities and the extension of Arnol'd's strange duality observed in [Ebeling-Wall]. For the space curve and surface singularities of Sections 3.4 and 3.6 we obtain Dynkin diagrams which are closely related to the Dynkin diagrams of Gabrielov for the unimodal hypersurface singularities. We study the Coxeter elements of these graphs and show how our results fit together with new results of K. Saito about Coxeter elements of a certain class of graphs. Section 3.6 provides supplementary information on the extension of the strange duality exceeding and completing the paper [Ebeling-Wall] in certain aspects. Finally we mention a particular result of our calculations: One finds topologically non-equivalent singularities, for example already among the triangle singularities, whose monodromy operators are conjugate over $\mathbb{Q}$ (Corollary 3.6.4).

The main results of this work are described in Chapter 4. These results are stated in Section 4.1. We first classify the isolated complete intersection singularities with a definite, parabolic or hyperbolic intersection form (Theorem 4.1.1). Then we give a description of the monodromy groups, of the relative monodromy groups, and of the sets of vanishing cycles and thimbles for almost all isolated singularities of even-dimensional complete intersections (Theorems 4.1.2 and 4.1.3). The only exceptions for which these characterizations are not true belong to the hyperbolic singularities. In Remark 4.1.4 we show that in this case one can find a description of these invariants in the framework of Kac-Moody-Lie algebras. For the hypersurface case and partially for the case of two-dimensional complete intersections in $\mathbb{C}^4$ these results are already published in [Ebeling_5] and generalize earlier results in [Pinkham_1], [Ebeling_1], [Ebeling_2], and [Ebeling_3]. At the end of Section 4.1 we quote the most important results corresponding to these results in the odd-dimensional case. Here we refer to [Janssen_1] for details. The proof of our central results consists in a reduction to algebraic theorems, which are proven in Chapter 5. For this reduction, which is described in Section 4.2, we need the results of the first three chapters. In Section 4.3 we discuss applications to global monodromy groups and Lefschetz pencils. In this way we return to the context in which the vanishing cycles and thimbles were originally introduced by S. Lefschetz.

In Chapter 5 we have collected the algebraic results on which the proof of the theorems in Chapter 4 is based. Here we consider subgroups Γ_Δ of the group of units of an integral symmetric lattice L which are generated by reflections corresponding to the vectors of a subset $\Delta \subset L$. Here the pair (L,Δ) has to satisfy the following conditions: (i) Δ consists only of minimal vectors of square length 2ε, $\varepsilon \in \{+1,-1\}$—fixed, (ii) Δ generates L, (iii) Δ is a Γ_Δ-orbit, (iv) unless $\operatorname{rk} L = 1$, there exist $\delta_1,\delta_2 \in \Delta$ with $\langle\delta_1,\delta_2\rangle = 1$. Such a pair is called a vanishing lattice, following a terminology of W.A.M. Janssen and E. Looijenga (Section 5.2). Typical examples of such vanishing lattices are the pairs (H,Δ) and $(\hat{H},\hat{\Delta})$. We show that a vanishing lattice which contains a certain small vanishing sublattice of Witt index 2 (and which is called complete in this case) is already the maximum possible vanishing lattice (Sections 5.3 and 5.4). This means that the subset Δ is maximum, hence contains all minimal vectors v of square length 2ε with $\langle v,L\rangle = \mathbb{Z}$, and Γ_Δ contains all reflections corresponding to minimal vectors of square length 2ε. It then follows from a theorem of M. Kneser [Kneser_1] that the elements of Γ_Δ are characterized by the properties that they have spinor norm 1 and act trivially on the quotient $L^\#/j(L)$ of the dual lattice $L^\#$ by the image of the lattice L. In Section 5.5 we show that these statements also hold true for some vanishing lattices defined by Coxeter systems, whereas in general these statements are not true for such vanishing lattices. Up to some supplements, Chapter 5 is largely identical with §§ 1 - 3 and part of § 5 of the paper [Ebeling_5].

This monograph is a translation of the author's "Habilitationsschrift" (Bonn 1986) with some minor modifications and corrections. A summary of the main results of the first four chapters is contained in the author's preprint "Vanishing lattices and monodromy groups of isolated complete intersection singularities" (to appear in Invent. math.).

This work was supported by a research grant of the Deutsche Forschungsgemeinschaft. I wish to express my thanks to this institution for this support. This grant also enabled me to visit the University of North Carolina at Chapel Hill/U.S.A. for two months in spring 1986 and to participate in the Special Year in Singularities and Algebraic Geometry. Part of this work was done during this period. I am especially grateful to the organizers of the Special Year, J. Damon and J. Wahl, for the hospitality and the pleasant working atmosphere.

I wish to thank all those who have contributed to this work by valuable suggestions and fruitful discussions, especially Lê Dũng Tràng, E. Looijenga, G. Pfister, K. Saito, J. Wahl, and above all C.T.C. Wall. I owe especial thanks to my teacher E. Brieskorn for his interest and his support and encouragement concerning this work. He has decisively influenced my way of thinking and viewing the problems.

The final preparation of the manuscript was supported by the "Max-Planck-Institut für Mathematik" in Bonn. I am especially grateful to Karin Deutler from this institute for preparing a beautiful camera-ready typescript.

Bonn, March 1987

Wolfgang Ebeling

TABLE OF CONTENTS

1. INVARIANTS OF COMPLETE INTERSECTIONS

1.1. The fundamental exact sequence

Let (X,x) be an isolated complete intersection singularity of dimension n. This means that (X,x) is a germ of a complex analytic space of dimension n, which is isomorphic to the fibre $(f^{-1}(0),0)$ of an analytic map-germ

$$f:(\mathbb{C}^{n+k},0) \longrightarrow (\mathbb{C}^{k},0),$$

and $x \in X$ is an isolated singular point of X. For $k=1$ (X,x) is an isolated hypersurface singularity.

An isolated complete intersection singularity has a semi-universal deformation

$$F:(\mathbb{C}^{n+p},0) \longrightarrow (\mathbb{C}^{p},0)$$

with smooth base space. Let B_ε be an open ball of radius ε around the origin in $\mathbb{C}^{n+p}$. We consider a representative of F of the form

$$F:X = F^{-1}(S) \cap B_\varepsilon \longrightarrow S$$

for a neighborhood S of 0 in $\mathbb{C}^p$. The germ of the set

$$C_F = \{y \in X \mid y \text{ is a critical point of } F\}$$

in 0 is the critical locus of F. The germ of its image $D_F = F(C_F)$ in 0 is the discriminant (locus) of F. The discriminant $(D_F,0)$ is a reduced irreducible hypersurface in $(\mathbb{C}^p,0)$.

For a sufficiently small ε and a sufficiently small neighborhood S of 0 in $\mathbb{C}^p$, the mapping

$$F|_{X-F^{-1}(D_F)}:X - F^{-1}(D_F) \longrightarrow S - D_F$$

is the projection of a differentiable fibre bundle. The typical fibre X_s of this bundle over a base point $s \in S - D_F$ has by [Milnor$_1$], [Hamm$_1$] the homotopy type of a bouquet of μ spheres S^n of real dimension n. The fibre X_s is called the Milnor fibre, the number μ the Milnor number of (X,x). The only non-trivial reduced homology group of X_s is the group $\tilde{H}_n(X_s,\mathbb{Z})$. This is a free $\mathbb{Z}$-module of rank μ.

The intersection number of cycles defines a bilinear form $\langle , \rangle$ on this module, which is symmetric for n even and skew-symmetric for n odd. We call

$$H = (\tilde{H}_m(X_s, \mathbb{Z}), \langle , \rangle)$$

the <u>Milnor lattice</u> of (X,x). If the dimension n is even, then one has in addition $\langle v,v \rangle \in 2\mathbb{Z}$ for all $v \in H$, i.e. the Milnor lattice is an even lattice in this case.

We now consider a construction, which can be traced back to Lê Dũng Tráng (cf. [Lê_2], [Looijenga_3]). We regard a generic complex line ℓ through the origin $0 \in \mathbb{C}^p$, i.e. a line which is not contained in the tangent cone of the discriminant $(D_F, 0)$ in 0. We assume that the coordinates of $\mathbb{C}^p$ are chosen in such a way that this line coincides with the last coordinate axis. This means in particular that this line meets the discriminant only in 0. But this is equivalent to the fact that

$$F' = (F_1, \ldots, F_{p-1}) : (\mathbb{C}^{n+p}, 0) \longrightarrow (\mathbb{C}^{p-1}, 0)$$

defines an isolated complete intersection singularity of dimension $n+1$.

We now choose the neighbourhood S of the form $S = T \times \bar{\mathbb{D}}$ for a disc

$$\bar{\mathbb{D}} = \{ z \in \mathbb{C} \mid |z| \leq \eta \}.$$

Let ε, η, and T be chosen so small and suitable that the following conditions are satisfied:

(i) For the representative

$$F' : X' = F'^{-1}(T) \cap B_\varepsilon \longrightarrow T,$$

the following is valid: The mapping

$$F'|_{X' - F'^{-1}(D_{F'})} : X' - F'^{-1}(D_{F'}) \longrightarrow T - D_{F'}$$

is the projection of a differentiable fibre bundle. The typical fibre of this bundle is a Milnor fibre of $(X', 0)$.

(ii) There exists a homeomorphism

$$h:\bar{X} = F^{-1}(S) \cap \bar{B}_\varepsilon \longrightarrow \bar{X}' = F'^{-1}(T) \cap \bar{B}_\varepsilon \ ,$$

such that the following diagram commutes:

$$\begin{array}{ccc} \bar{X} & \xrightarrow{\quad F \quad} & S = T \times \bar{\mathbb{D}} \\ \Big\downarrow h & & \Big\downarrow \pi \\ \bar{X}' & \xrightarrow{\quad F' \quad} & T \end{array} \quad .$$

Here π denotes the projection onto the first factor, and h is the identity on $F^{-1}(T \times \mathbb{D}_1)$ with $\bar{\mathbb{D}}_1 \subset \mathbb{D}$. This is possible by [Looijenga$_3$, Proposition (5.4)].

(iii) $S - D_F$ represents the homotopy type of $S - D_F$ at 0 [Looijenga$_3$, (7.3)].

The restriction of the projection $\pi:S \longrightarrow T$ to the discriminant $D = D_F$ is finite. Let t be a point of T which is not contained in the image of the ramification locus of $\pi|_D$.

We now consider the mapping

$$F|_{X_t'}:X_t' = F'^{-1}(t) \cap B_\varepsilon \longrightarrow \{t\} \times \bar{\mathbb{D}} \ .$$

This mapping corresponds to the function

$$F_p:X_t' \longrightarrow \bar{\mathbb{D}} \ .$$

By the choice of t there are exactly m intersection points of $\{t\} \times \bar{\mathbb{D}}$ with the discriminant D, where m is the multiplicity of the discriminant at the origin. We denote these points by $s_1,\dots,s_m$. A fibre X_{s_i} over such a point has exactly one singularity, and this is an ordinary double point. Off these singular fibres, the mapping $F|_{X_t'} = F_p$ is the projection of a differentiable fibre bundle. We choose $s = (t,\eta) \in \{t\} \times \bar{\mathbb{D}}$ as base point. The fibre X_s over the point s is a Milnor fibre of (X,x). The situation is illustrated in Figure 1.1.1. Here the manifold X_t' appears as a ball, which is, however, in fact only true in the hypersurface case. The manifold X_t' has the homotopy type of a bouquet of μ' spheres S^{n+1}, where μ'

is the Milnor number of the singularity $(X',0)$.

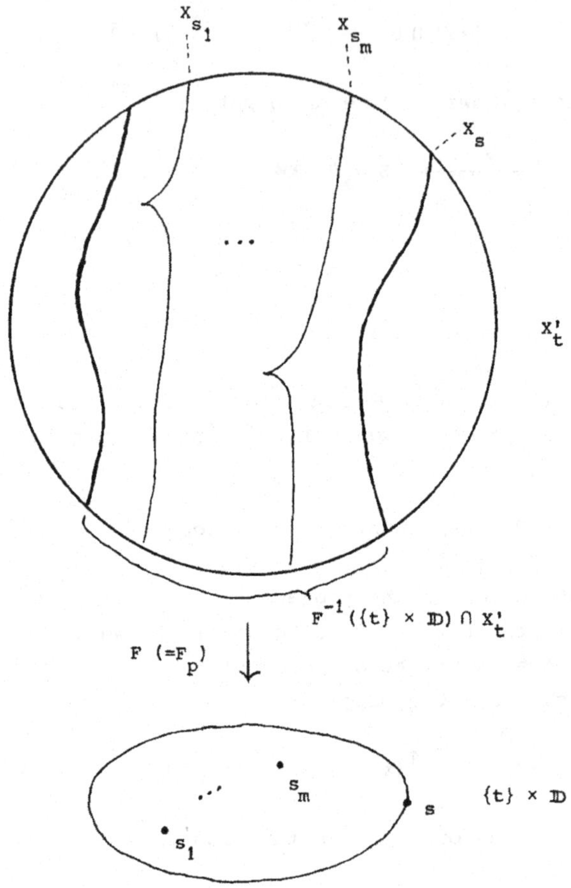

Figure 1.1.1

The exact (reduced) homology sequence of the pair (X'_t, X_s) therefore reduces to the following short exact sequence

$$0 \longrightarrow H_{n+1}(X'_t) \longrightarrow H_{n+1}(X'_t, X_s) \xrightarrow{\partial_*} \tilde{H}_n(X_s) \longrightarrow 0 \qquad (*)$$

where the possibly non-trivial terms are free $\mathbb{Z}$-modules of rank μ', m, respectively μ. In particular

$$m = \mu' + \mu \quad .$$

We define on $H_{n+1}(X_t', X_s)$ a bilinear form, also denoted by $\langle , \rangle$, by the pull-back of the bilinear form $\langle , \rangle$ on H via the mapping ∂_*. We denote the corresponding lattice by $\hat{H}$. Therefore

$$\langle \hat{v}, \hat{w} \rangle := \langle \partial_* \hat{v}, \partial_* \hat{w} \rangle$$

for $\hat{v}, \hat{w} \in \hat{H}$. The motivation for this definition are the Picard-Lefschetz formulas (see Section 1.3). This definition is justified by the fact that $\hat{v}$ and $\hat{w}$ can be represented by relative cycles which intersect only on X_s (see Remark 1.4.1).

On $H_{n+1}(X_t')$ we consider the bilinear form induced by inclusion in $\hat{H}$ (ignoring the intersection form on this module). We denote the corresponding lattice by H'. By the definition of the bilinear form on $\hat{H}$, H' is contained in the kernel (radical) of $\hat{H}$, which we denote by $\ker \hat{H}$. With these notations the above exact sequence (*) can be reformulated as follows:

$$0 \longrightarrow H' \longrightarrow \hat{H} \xrightarrow{\partial_*} H \longrightarrow 0 \quad .$$

In the hypersurface case H' is trivial and ∂_* is an isomorphism of $\hat{H}$ onto H. In this case we identify $\hat{H}$ and H. In the general case this fundamental exact sequence replaces the single invariant H in the hypersurface case. The invariants introduced below will be connected with this sequence.

1.2. Vanishing cycles and thimbles

We now consider the local situation around the points $s_i \in \bar{D}$. Let x_i be the singular point of the fibre X_{s_i}. This is an ordinary double point. Therefore there exists a small neighbourhood B_i of x_i in X_t' and local coordinates $(u_1, \ldots, u_{n+1})$ in this neighbourhood, such that $F|_{B_i}$ can be written in these coordinates as follows

$$F(u_1, \ldots, u_{n+1}) = s_i + u_1^2 + \ldots + u_{n+1}^2 \ ,$$

and B_i is a ball of radius ε' in these coordinates. Moreover, we choose a small disc $\bar{D}_i$ centered at s_i with radius ρ, and set

$$Y_i = F^{-1}(\bar{D}_i) \cap B_i, \quad Y_*^{(i)} = F^{-1}(s_i + \rho) \cap B_i \quad .$$

Here ρ is chosen so small that

$$F|_{Y_i} : Y_i \longrightarrow \mathbb{D}_i$$

is a Milnor fibration outside s_i. Moreover, let the neighbourhoods B_i and the discs $\bar{\mathbb{D}}_i$ be chosen so small that they are contained in X'_t respectively $\mathbb{D}$.

The sets $Y = Y_i$ and $Y_* = Y_*^{(i)}$ can be described in the above coordinates as follows:

$$Y = \{u \in \mathbb{C}^{n+1} \mid |u_1|^2 + \ldots + |u_{n+1}|^2 \leq \varepsilon'^2 \text{ and } |u_1^2 + \ldots + u_{n+1}^2| \leq \rho\},$$

$$Y_* = \{u \in Y \mid u_1^2 + \ldots + u_{n+1}^2 = \rho\}.$$

The description of Y shows that Y can be linearly contracted onto the origin. The fundamental exact sequence for an ordinary double point therefore reduces to

$$H_{n+1}(Y, Y_*) \xrightarrow{\partial_*} H_n(Y_*),$$

where the connecting homomorphism ∂_* is an isomorphism.

In particular Y can be retracted onto the real $n+1$ - disc

$$D^{n+1} = \{u \in Y \mid \text{all } u_\nu \text{ real}\}.$$

One can easily show (see e.g. [Lamotke$_1$, 5.5]) that the fibre Y_* has the boundary of this disc, the real sphere

$$S^n = \partial D^{n+1} = \{u \in Y_* \mid \text{all } u_\nu \text{ real}\},$$

as a strong deformation retract. After the choice of an orientation of D^{n+1} and the orientation of S^n as the boundary of D^{n+1}, the homology classes $[D^{n+1}]$, respectively $[S^n]$, yield generators of $H_{n+1}(Y, Y_*)$, respectively $H_n(Y_*)$, with $\partial_*[D^{n+1}] = [S^n]$.

Now let $\varphi = \varphi_i : [0,1] \longrightarrow \bar{\mathbb{D}}$ be a path from s_i to s with $\varphi((0,1]) \subset \mathbb{D}^* = \bar{\mathbb{D}} - \{s_1, \ldots, s_m\}$ meeting the boundary of the disc $\bar{\mathbb{D}}_i$ precisely in $s_i + \rho$ at time $0 < \Theta < 1$. Therefore let $\varphi(\Theta) = s_i + \rho$. Let $\psi = \varphi([\Theta, 1])$. Since the inverse image $(F|_{X'_t})^{-1}(\psi)$ of ψ is fibered trivially, there is an embedding

$$j : Y_* \times \psi \longrightarrow X'_t$$

with $j(Y_* \times \psi) = (F|_{X_t'})^{-1}(\psi)$, $j(y, s_i + \rho) = y$ and $(F \circ j)(y,\lambda) = \lambda$ for $y \in Y_*$ and $\lambda \in \psi$. Then

$$Z = D^{n+1} \cup j(S^n \times \psi)$$

(with the chosen orientation) defines an element

$$\hat{\delta} \in \hat{H} = H_{n+1}(X_t', X_s).$$

The boundary ∂Z of Z defines an element

$$\delta \in H = H_n(X_s)$$

with $\partial_* \hat{\delta} = \delta$ (cf. Figure 1.2.1). Following Lefschetz [Lefschetz$_1$] we call δ a <u>vanishing cycle</u> (corresponding to the path φ). Lefschetz denotes $\hat{\delta}$ in [Lefschetz$_1$] by the French word "onglet". The usual notation for $\hat{\delta}$ in English is "<u>thimble</u>", which corresponds to the German word "Fingerhut".

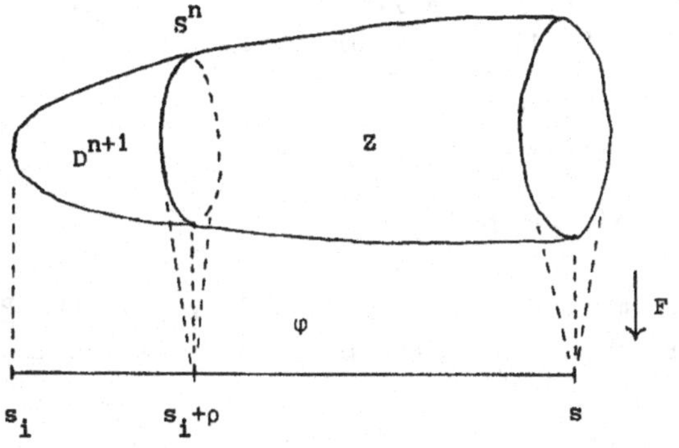

<u>Figure 1.2.1.</u>

The normal bundle of S^n in Y_* is isomorphic to the tangent bundle of the sphere S^n. This yields

$$\langle\hat{\delta},\hat{\delta}\rangle = \langle\delta,\delta\rangle = \begin{cases} 0 & \text{if } n \text{ is odd,} \\ (-1)^{n/2}2 & \text{if } n \text{ is even.} \end{cases}$$

1.3. Monodromy groups and vanishing lattices

In this section we study the action of the fundamental group $\pi_1(\mathbb{D}^*,s)$ on H and $\hat{H}$, where $\mathbb{D}^* = D - \{s_1,\ldots,s_m\}$.

With the path φ a simple loop $\omega = \omega_i$ is associated as follows. Let

$$\tilde{\omega}_i(r) = s_i + \rho e^{2\pi\sqrt{-1}r} , \qquad 0 \le r \le 1 ,$$

be the path going once around s_i along the boundary of $\bar{\mathbb{D}}_i$. Then define $\omega = \omega_i : [0,1] \longrightarrow X_s$ by

$$\omega = \psi\tilde{\omega}_i\psi^{-1} .$$

The loop ω now induces automorphisms $\gamma \in \mathrm{Aut}(H)$ and $\hat{\gamma} \in \mathrm{Aut}(\hat{H})$ as follows. Since the pull-back of the bundle $F_p : F_p^{-1}(\mathbb{D}^*) \cap X_t' \longrightarrow \mathbb{D}^*$ to the interval $[0,1]$ via ω is trivial, there is a continuous mapping

$$W : X_s \times [0,1] \longrightarrow X_t'$$

with $(F_p \circ W)(y,r) = \omega(r)$ and $W(y,0) = y$ for $y \in X_s$, $r \in [0,1]$, inducing for each $r \in [0,1]$ a homeomorphism

$$W_r : X_s \xrightarrow{\cong} X_{\omega(r)}$$

$$y \longmapsto W(y,r) .$$

Then γ is the automorphism of H induced by W_1, the Picard-Lefschetz transformation corresponding to the path φ. The mapping W induces a homomorphism

$$W_* : H_{n+1}(X_s \times [0,1],\ X_s \times \{0\} \cup X_s \times \{1\}) \longrightarrow H_{n+1}(X_t', X_s)$$

of relative homology groups. The composition of the mapping

$$H_n(X_s) \longrightarrow H_{n+1}(X_s \times [0,1],\ X_s \times \{0\} \cup X_s \times \{1\}) ,$$

$$y \longrightarrow y \times \iota ,$$

where $\iota \in H_1([0,1],\{0,1\})$ is the canonical generator, with W_* yields a homomorphism

$$\tau_\omega : H \longrightarrow \hat{H} ,$$

which we call the <u>extension along the path</u> ω (cf. [Lamotke$_1$, (6.4)], where also important properties of τ_ω are stated). We now define (cf. [Dimca$_1$, §2]):

$$\hat{\gamma}(y) = y + (-1)^n \tau_\omega(\partial_* y).$$

This is an automorphism of $\hat{H}$, which depends like γ only on the homotopy class of the path ω. We call $\hat{\gamma}$ the <u>relative Picard-Lefschetz transformation</u> corresponding to φ.

The action of γ, respectively $\hat{\gamma}$, on H, respectively $\hat{H}$, is described by the <u>Picard-Lefschetz formulas</u>. For $y \in H$ one has

$$\gamma(y) = y - (-1)^{n(n-1)/2} \langle y, \delta \rangle \delta ,$$

and for $\hat{y} \in \hat{H}$:

$$\hat{\gamma}(\hat{y}) = \hat{y} - (-1)^{n(n-1)/2} \langle \partial_* \hat{y}, \partial_* \hat{\delta} \rangle \hat{\delta}$$
$$= \hat{y} - (-1)^{n(n-1)/2} \langle \hat{y}, \hat{\delta} \rangle \hat{\delta} .$$

The first formula is the usual Picard-Lefschetz formula, the second one follows from [Lamotke$_1$, (6.7.1)]. Because of

$$\gamma(y) = y + (-1)^n \partial_* \tau_\omega(y) \qquad \text{for} \quad y \in H$$

([Lamotke$_1$, (6.4.6)]), one has

$$\partial_* \hat{\gamma} = \gamma \partial_* .$$

Therefore the two formulas are equivalent to each other.

Let $L^{(n)}$ be a lattice which is symmetric for n even and skew-symmetric for n odd. We define for $\lambda \in L^{(n)}$ with

$$\langle \lambda, \lambda \rangle = \begin{cases} 0 & \text{for } n \text{ odd,} \\ (-1)^{n/2} 2 & \text{for } n \text{ even,} \end{cases}$$

an automorphism $s_\lambda^{(n)} \in \mathrm{Aut}(L^{(n)})$ by

$$s_\lambda^{(n)}(y) = y - (-1)^{n(n-1)/2} \langle y,\lambda\rangle \lambda \quad \text{for} \quad y \in L^{(n)} .$$

For n even, $s_\lambda^{(n)}$ is a reflection, and for n odd a symplectic transvection. With this notation the Picard-Lefschetz formulas can be reformulated as follows:

$$\gamma = s_\delta^{(n)} \quad , \quad \hat{\gamma} = s_{\hat{\delta}}^{(n)} \quad .$$

The formula for $\hat{\gamma}$ is the motivation for the definition of the bilinear form on $\hat{H}$.

We have therefore defined a representation of the fundamental group $\pi_1(\mathbb{D}^*,s)$ on H and $\hat{H}$. We denote the images of these representations in $\mathrm{Aut}(H)$, respectively $\mathrm{Aut}(\hat{H})$, by Γ, respectively $\hat{\Gamma}$. Since the inclusion $j:\mathbb{D}^* \longrightarrow S - D_F$ induces a surjective homomorphism of the corresponding fundamental groups [Looijenga$_3$, (7.1)], Γ coincides with the monodromy group of the singularity (X,x) (bear condition (iii) of Section 1.1 in mind). The monodromy group of (X,x) is the image of the representation of $\pi_1(S - D_F,s)$ on H. We call the group $\hat{\Gamma}$ the relative monodromy group of (X,x).

If one admits instead of a path φ contained in $\{t\} \times \bar{\mathbb{D}}$ an arbitrary path from a regular point of the discriminant D_F to s, then the above construction also yields a vanishing cycle $\pm\delta \in H$ (determined up to sign). The set of all these homology classes is called the set of vanishing cycles, and will be denoted by $\Delta \subset H$. It is well known (see e.g. [Looijenga$_3$, (7.8)]) that the set Δ is one orbit under the action of the monodromy group Γ , except in the case of an ordinary double point A_1. Therefore the set Δ coincides with the set of all vanishing cycles defined by paths entirely contained in $\{t\} \times \mathbb{D}$.

The set of thimbles $\hat{\Delta}$ is defined to be the set of all thimbles $\hat{\delta}$ defined by any path φ in $\{t\} \times \mathbb{D}$ to any of the points s_i as above. Analogously this is a $\hat{\Gamma}$-orbit, again excluding the case A_1.

Let $L^{(n)}$ be a symmetric or skew-symmetric lattice as above and let $\Lambda^{(n)}$ be a subset of vectors λ with

$$\langle\lambda,\lambda\rangle = \begin{cases} 0 & \text{for } n \text{ odd,} \\ (-1)^{n/2}2 & \text{for } n \text{ even,} \end{cases}$$

and $s_\lambda^{(n)}$ the automorphisms of $L^{(n)}$ defined as above. Let $\Gamma_{\Lambda}(n)$ be the subgroup of $\mathrm{Aut}(L^{(n)})$ generated by the $s_\lambda^{(n)}$ for $\lambda \in \Lambda(n)$.

A pair $(L^{(n)}, \Lambda^{(n)})$ is called a vanishing lattice, if the following conditions are satisfied (cf. [Looijenga$_3$, (7.9)], [Janssen$_1$]):

(i) $\Lambda^{(n)}$ generates the lattice $L^{(n)}$.

(ii) $\Lambda^{(n)}$ is one orbit under $\Gamma_{\Lambda(n)}$.

(iii) If $\operatorname{rk} L^{(n)} > 1$, then there exist $\lambda_1, \lambda_2 \in \Lambda^{(n)}$ with $\langle\lambda_1, \lambda_2\rangle = 1$.

The group $\Gamma_{\Lambda(n)}$ is called the corresponding monodromy group.

Then not only (H, Δ) but also $(\hat{H}, \hat{\Delta})$ is a vanishing lattice, and the group Γ, respectively $\hat{\Gamma}$, is its corresponding monodromy group. Condition (iii) is implied by the fact that each singularity, which is not of type A_1, deforms to a singularity of type A_2 [Looijenga$_3$, (7.18)].

1.4. Geometric bases

We consider again the situation of Section 1.2, and we choose an ordered system of paths $(\varphi_1, \ldots, \varphi_m)$ from the points $s_1, \ldots, s_m$ to s. Let the neighbourhoods B_i and the radius ρ of the discs $\mathbb{D}_i$ be chosen in such a way that all these sets are disjoint.

We call the system of paths weakly distinguished, if the simple loops ω_i corresponding to the paths φ_i form a free system of generators of $\pi_1(\mathbb{D}^*, s)$.

We call the system of paths (strongly) distinguished (or geometric), if the following conditions are satisfied:

(i) The paths φ_i are non-self-intersecting.

(ii) The only common point of φ_i and φ_j for $i \neq j$ is s.

(iii) The paths are numbered according to the order in which they arrive at s, counted clockwise beginning from the boundary of the disc (see Figure 1.4.1, cf. [Husein-Zade$_1$, 1.2]).

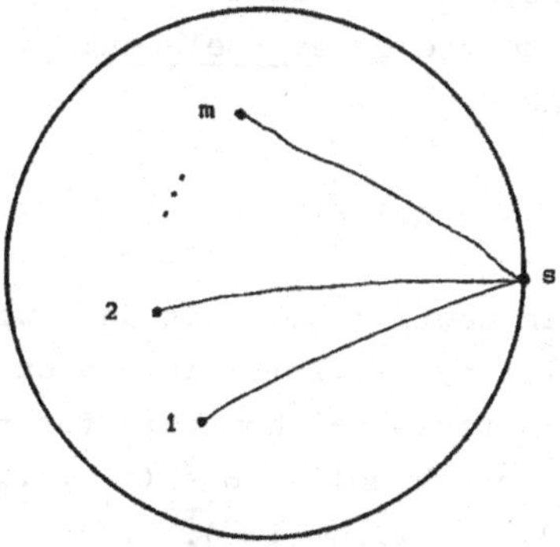

Figure 1.4.1.

A strongly distinguished system of paths is of course also weakly distinguished.

An ordered system of paths $(\varphi_1,\ldots,\varphi_m)$ defines according to Section 1.2 an ordered system of vanishing cycles $B = (\delta_1,\ldots,\delta_m)$ and an ordered system of thimbles $\hat{B} = (\hat{\delta}_1,\ldots,\hat{\delta}_m)$, which we call <u>weakly</u>, respectively <u>strongly distinguished</u>, if the system of paths has these properties. With similar methods as in [Brieskorn$_1$, Appendix] (see also [Looijenga$_3$, 5.B]) one can show that a strongly distinguished system of thimbles $\hat{B}$ forms a basis of $\hat{H}$. By (Husein-Zade$_1$, 2.1] this follows for a weakly distinguished system, too. We call $\hat{B}$ a <u>strongly</u>, respectively a <u>weakly, distinguished basis of thimbles</u>.

The ordered system $B = (\delta_1,\ldots,\delta_m)$ of vanishing cycles corresponding to a strongly or weakly distinguished basis of thimbles forms a system of generators of H, which we call <u>strongly</u>, respectively <u>weakly distinguished</u>, too.

The importance of these bases (and corresponding systems of generators) is explained by the following relations. First let $\hat{B} = (\hat{\delta}_1,\ldots,\hat{\delta}_m)$ be a weakly distinguished basis of thimbles, and let $B = (\delta_1,\ldots,\delta_m)$ be the corresponding weakly distinguished system of generators of vanishing cycles. Then the corresponding Picard-Lefschetz transformations

$$\hat{\gamma}_i = s^{(n)}_{\hat{\delta}_i}\text{, respectively}\quad \gamma_i = s^{(n)}_{\delta_i},\quad \text{for}\quad i = 1,\ldots,m\ ,$$

generate the relative monodromy group $\hat{\Gamma}$, respectively the monodromy group Γ.

Running around the disc $\mathbb{D}$ along the boundary $\partial\mathbb{D}$ in the positive direction (counterclockwise) induces an automorphism $\hat{c} \in \hat{\Gamma}$ as well as an automorphism $c \in \Gamma$ according to Section 1.3. We call c a (generic) <u>monodromy operator</u> and $\hat{c}$ a (generic) <u>relative monodromy operator</u> of (X,x). Now if $\hat{B}$ and B are strongly distinguished, then $\hat{c}$, respectively c, correspond to the <u>Coxeter elements</u> with respect to $\hat{B}$, respectively B, i.e. one has

$$\hat{c} = \hat{\gamma}_1\hat{\gamma}_2 \cdots \hat{\gamma}_m\ ,\ \text{respectively}\quad c = \gamma_1\gamma_2 \cdots \gamma_m.$$

<u>Remark 1.4.1.</u> We mentioned in Section 1.1 that any two elements of $\hat{H}$ can be represented by relative cycles which intersect only in X_s. This can be seen as follows. It suffices to show this for the elements of a strongly distinguished basis of thimbles $\hat{B} = (\hat{\delta}_1,\ldots,\hat{\delta}_m)$. This is true by construction for $\hat{\delta}_i$ and $\hat{\delta}_j$ with $i \neq j$. Since y is linearly

contractible to the origin, the class $[D^{n+1}] \in H_{n+1}(Y, Y_*)$ used in Section 1.2 for the construction of the thimble $\hat{\delta}_i$ can also be represented by a disc $\tilde{D}^{n+1}$ with

$$\tilde{D}^{n+1} \cap D^{n+1} = \partial\tilde{D}^{n+1} = \partial D^{n+1} = S^n \subset Y_* .$$

From this one can easily derive the claim.

Remark 1.4.2. It is not known whether one can always select a basis of the Milnor lattice H from a strongly or weakly distinguished system of generators of vanishing cycles. We shall give such systems of generators for which this is possible for the special singularities considered in this book (cf. Chapter 3). M. Merle, however, has recently informed me that he can show that the Milnor lattice H always has a basis consisting of vanishing cycles [Merle$_1$]. The methods which he uses are connected with Chapter 2.1.

1.5. Dynkin diagrams

Let $\Lambda = (\lambda_1, \ldots, \lambda_r)$ be an ordered system of thimbles $\lambda_i \in \hat{\Delta}$ or of vanishing cycles $\lambda_i \in \Delta$ for $i = 1, \ldots, r$. The matrix (intersection matrix) $((\langle \lambda_i, \lambda_j \rangle))$ describing the bilinear form $\langle , \rangle$ with respect to Λ is usually represented by a graph D_Λ as follows. The vertices $e_1, \ldots, e_r$ of D_Λ correspond to the elements $\lambda_1, \ldots, \lambda_r$, and for $i < j$ the vertex e_i is connected with e_j by $|\langle \lambda_i, \lambda_j \rangle|$ edges weighted by the sign $+1$ or -1 of $\langle \lambda_i, \lambda_j \rangle \in \mathbb{Z}$. We indicate the weight

$$\varepsilon = \begin{cases} (-1)^{n/2} & \text{for } n \text{ even,} \\ (-1)^{n+1/2} & \text{for } n \text{ odd} \end{cases}$$

in the figures of the graphs by a dotted line, the weight $-\varepsilon$ by a usual line. This graph D_Λ is called the (Coxeter-)Dynkin diagram corresponding to Λ.

The Dynkin diagram $D_{\hat{B}}$ corresponding to a basis $\hat{B}$ of thimbles coincides with the Dynkin diagram D_B corresponding to the associated system of generators B of vanishing cycles according to the definition of the bilinear form on $\hat{H}$. We define:

<u>Definition 1.5.1.</u> The set of all Dynkin diagrams corresponding to strongly (respectively weakly) distinguished bases of thimbles is denoted by $\mathcal{D}^*$ (respectively $\mathcal{D}^0$).

<u>Remark 1.5.2.</u> The graphs of $\mathcal{D}^*$ and $\mathcal{D}^0$ are connected (see [Looijenga$_3$,(7.5)]).

The definition of a strongly (weakly) distinguished m-tuple $(\lambda_1,\dots,\lambda_m)$ of thimbles or vanishing cycles depends on several choices, such as the choice of the line ℓ, of the points t and s, of the paths φ_i, and of the orientations of the λ_i. We now consider actions on the set of all such m-tuples and therefore on the set of the corresponding Dynkin diagrams (cf. Gabrielov$_1$],[Husein-Zade$_1$],[Ebeling$_4$],[Brieskorn$_3$],[Slodowy$_1$]).

(a) Action of $(\mathbb{Z}/2\mathbb{Z})^m$ (change of orientation):

$$\kappa_j(\lambda_1,\dots,\lambda_m) = (\lambda_1,\dots,\lambda_{j-1},-\lambda_j,\lambda_{j+1},\dots,\lambda_m), \quad j = 1,\dots,m$$

(b) Action of Γ (respectively $\hat{\Gamma}$):

$$\gamma(\lambda_1,\dots,\lambda_m) = (\gamma(\lambda_1),\dots,\gamma(\lambda_m)),$$

where $\gamma \in \Gamma$ if $\lambda_i \in \Delta$, $\gamma \in \hat{\Gamma}$ if $\lambda_i \in \hat{\Delta}$ for all i.

(c) Action of the braid group Z_m:
Let Z_m be the braid group with m strings, and let $\alpha_1,\dots,\alpha_{m-1}$ be its standard generators. For $j = 1,\dots,m-1$ the generator α_j acts as follows:

$$\alpha_j(\lambda_1,\dots,\lambda_m) = (\lambda_1,\dots,\lambda_{j-1},s^{(n)}_{\lambda_j}(\lambda_{j+1}),\lambda_j,\lambda_{j+2},\dots,\lambda_m).$$

The inverse transformation α_j^{-1} is also denoted by β_{j+1}.

(d) Action of the symmetric group S_m:

$$\sigma(\lambda_1,\dots,\lambda_m) = (\lambda_{\sigma(1)},\dots,\lambda_{\sigma(m)}), \ \sigma \in S_m$$

(e) "Gabrielov transformations"

$$\alpha_i(j)(\lambda_1,\dots,\lambda_m) = (\lambda_1,\dots,\lambda_{j-1},s^{(n)}_{\lambda_i}(\lambda_j),\lambda_{j+1},\dots,\lambda_m),$$

$$\beta_i(j) = (\alpha_i(j))^{-1}, \quad i,j \in \{1,\dots,m\}.$$

All these actions retain the property of an m-tuple $(\lambda_1,\ldots,\lambda_m)$ to be a system of generators or a basis. The transformations (c) are induced by the following transformations on the system of paths $(\varphi_1,\ldots,\varphi_m)$, respectively on the system of the corresponding simple loops $(\omega_1,\ldots,\omega_m)$:

$$\alpha_j(\varphi_1,\ldots,\varphi_m) = (\varphi_1,\ldots,\varphi_{j-1},\varphi_{j+1}\omega_j,\varphi_j,\varphi_{j+2},\ldots,\varphi_m) \;,$$

$$\alpha_j(\omega_1,\ldots,\omega_m) = (\omega_1,\ldots,\omega_{j-1},\omega_j\omega_{j+1}\omega_j^{-1},\omega_j,\omega_{j+2},\ldots,\omega_m) .$$

The transformations (e) were introduced in this context by Gabrielov [Gabrielov$_1$], and are combinations of transformations of (c) and (d). The transformations $\alpha_i(j)$ and $\beta_i(j)$ act on the set $\mathcal{D}^0$ as follows. For

$$(\lambda_1,\ldots,\lambda_m) \underset{\beta_i(j)}{\overset{\alpha_i(j)}{\rightleftarrows}} (\tilde{\lambda}_1,\ldots,\tilde{\lambda}_m)$$

one has:

$$\langle\tilde{\lambda}_r,\tilde{\lambda}_s\rangle = \langle\lambda_r,\lambda_s\rangle \quad \text{for} \quad 1 \leq r,\ s \leq m,\ r,s \neq j \quad \text{or} \quad r = s = j,$$

$$\langle\tilde{\lambda}_i,\tilde{\lambda}_j\rangle = (-1)^{n+1}\langle\lambda_i,\lambda_j\rangle \quad \text{for} \quad i \neq j \;,$$

$$\left.\begin{aligned} \langle\tilde{\lambda}_r,\tilde{\lambda}_j\rangle &= \langle\lambda_r,\lambda_j\rangle - (-1)^{n(n-1)/2}\langle\lambda_r,\lambda_i\rangle\langle\lambda_j,\lambda_i\rangle \\ \langle\lambda_r,\lambda_j\rangle &= \langle\tilde{\lambda}_r,\tilde{\lambda}_j\rangle - (-1)^{n(n+1)/2}\langle\tilde{\lambda}_r,\tilde{\lambda}_i\rangle\langle\tilde{\lambda}_j,\tilde{\lambda}_i\rangle \end{aligned}\right\} \text{for} \quad r \neq i,j.$$

From this one can easily deduce the action (c) on the set $\mathcal{D}^*$ because of the following relations between the actions (c), (d), and (e). If $\tau_{j,j+1} \in S_m$ denotes the transposition of j and $j+1$, one has:

$$\alpha_j = \tau_{j,j+1} \circ \alpha_j(j+1) \quad ,$$

$$\beta_{j+1} = \tau_{j,j+1} \circ \beta_{j+1}(j) \quad .$$

The action (b) serves to compensate the indeterminacy of the base point $s \in S - D_F$, and is trivial on the sets $\mathcal{D}^0$ and $\mathcal{D}^*$. All the transformations leave the (relative) monodromy group Γ (respectively $\hat{\Gamma}$) invariant, the transformations (d) and (e), however, can change the conjugacy class of c and $\hat{c}$ (compare the examples in [Ebeling_1],[Ebeling_4]).

Definition 1.5.3. We call two Dynkin diagrams D and D' corresponding to bases of thimbles weakly equivalent, if they can be transformed into each other by iteration of transformations of the classes (a), (c), (d), and (e). We call them strongly equivalent, if they can be transformed into each other by iteration of transformations (a) and (c).

Proposition 1.5.4. Any two Dynkin diagrams of $\mathcal{D}^*$ are strongly equivalent.

This proposition was proven by Gabrielov, see [Husein-Zade_1,2.2.3].

Proposition 1.5.5 (S. Humphries). Any two Dynkin diagrams of $\mathcal{D}^0$ are weakly equivalent.

This was conjectured by Husein-Zade [Husein-Zade_1, 2.2.5] and proven by S. Humphries [Humphries_1]. The proof of Humphries is elementary. As was observed by R. Pellikaan, the proposition also follows from [Lyndon-Schupp, Proposition 4.20].

In order to show that the sets $\mathcal{D}^*$ and $\mathcal{D}^0$ are invariants of the singularity (X,x), we still have to verify their independence of the choice of the line ℓ and of the point $t \in T$. The independence of the choice of the point t for a fixed line ℓ follows as in [Siersma_1, §7] (see also [Looijenga_3,(7.13)]), since the image V_ℓ of the ramification locus of $\pi|_D = \pi_\ell|_D$ in $T = T_\ell$ is a proper analytic subset of T_ℓ, hence the complement $T_\ell - V_\ell$ is open, dense, and pathwise connected in T_ℓ. The set $\mathcal{T}_0(D)$ of all lines contained in the tangent cone of $(D,0)$ is a proper analytic subset of $\mathbb{P}^{p-1}$, the set of all lines in $\mathbb{C}^p$ through the origin. The complement is therefore also open, dense, and pathwise connected. Let $\ell,\ell' \in \mathbb{P}^{p-1} - \mathcal{T}_0(D)$ and $t \in T_\ell - V_\ell$, $t' \in T_{\ell'} - V_{\ell'}$. Then $\{t\} \times \ell$ and $\{t'\} \times \ell'$ can be joined by a piecewise smooth path in $\mathbb{C}^p \times \mathbb{P}^{p-1}$ avoiding the sets $\mathbb{C}^p \times \mathcal{T}_0(D)$ and

$$\bigcup_{\ell \in \mathbb{P}^{p-1} - \mathcal{T}_0(D)} V_\ell \times \{\ell\}.$$

The independence of $\mathcal{D}^0$ and $\mathcal{D}^*$ from the choice of ℓ and t then follows similarly as in [Dimca_1, §1] with the help of simple stratification arguments from the Second Isotopy Lemma of Thom-Mather [Gibson et al., II (5.8)].

Moreover, the same arguments imply the independence of the fundamental exact sequence

$$0 \longrightarrow H' \longrightarrow \hat{H} \longrightarrow H \longrightarrow 0$$

of the choice of the generic line ℓ. The invariance of $\mathcal{D}^*$ also implies the invariance of the conjugacy class $\hat{C}^* \subset \hat{\Gamma}$ of the relative monodromy operator $\hat{c}$ and, using the independence of the above sequence, the invariance of the conjugacy class $C^* \subset \Gamma$ of the generic monodromy operator c.

We have therefore shown that the introduced objects

$$H', \hat{H}, H; \hat{\Delta}, \Delta; \hat{\Gamma}, \Gamma; \mathcal{D}^*, \mathcal{D}^0; \hat{C}^*, C^*$$

are all invariants of the singularity (X,x).

1.6. Relative monodromy

In the hypersurface case the invariant $\mathcal{D}^*$, respectively already a single Dynkin diagram of $\mathcal{D}^*$, determines the Milnor lattice H, the set of vanishing cycles $\Delta \subset H$, the monodromy group $\Gamma \subset \mathrm{Aut}(H)$, and the conjugacy class C^* of the monodromy operator $c \in \Gamma$. (See the article of Brieskorn [Brieskorn$_3$] for a survey of the relations between the various invariants in the hypersurface case.) If, however, (X,x) is not a hypersurface singularity, then $\mathcal{D}^*$, respectively $D \in \mathcal{D}^*$, determines a priori only the relative objects $\hat{H}, \hat{\Delta}, \hat{\Gamma}$, and $\hat{C}^*$. For the knowledge of H, Δ, Γ, and C^* one needs additional information on the linear relations between the vanishing cycles $\delta_1, \ldots, \delta_m$ of a strongly distinguished system of generators corresponding to the Dynkin diagram D, i.e. on the module H'. We discuss in this section to what extent this information can be obtained from the relative monodromy, the Coxeter element $\hat{c}$ corresponding to D.

Let $(\hat{\delta}_1, \ldots, \hat{\delta}_m) \in \hat{\Delta}^m$ be a strongly distinguished basis of thimbles with corresponding relative monodromy operator $\hat{c} \in \hat{\Gamma}$ and corresponding monodromy operator $c \in \Gamma$. By Sections 1.1 and 1.3 one has the following commutative diagram with exact rows:

$$\begin{array}{ccccccccc} 0 & \longrightarrow & H' & \longrightarrow & \hat{H} & \xrightarrow{\partial_*} & H & \longrightarrow & 0 \\ & & \downarrow{\scriptstyle 0} & & \downarrow{\scriptstyle \hat{c}-id_{\hat{H}}} & & \downarrow{\scriptstyle c-id_H} & & \\ 0 & \longrightarrow & H' & \longrightarrow & \hat{H} & \xrightarrow{\partial_*} & H & \longrightarrow & 0 \end{array} \qquad (1.6.1)$$

Here the first vertical homomorphism from the left is the zero homomorphism. By the Snake Lemma [Bourbaki$_1$, Chap.1, § 1.4] one can deduce the following exact sequence:

$$0 \to H' \to \ker(\hat{c}-id_{\hat{H}}) \to \ker(c-id_H) \to H' \to \operatorname{coker}(\hat{c}-id_{\hat{H}}) \to \operatorname{coker}(c-id_H) \to 0 \qquad (1.6.2)$$

In particular one has

$$H' \subset \ker(\hat{c} - id_{\hat{H}}) \ .$$

In certain cases one can determine this submodule more precisely. Therefore we need the following preparations.

We write the intersection matrix $A = ((\langle\hat{\delta}_i,\hat{\delta}_j\rangle))$ in the form $A = V + (-1)^n V^t$, where V is an upper triangular matrix with $(-1)^{n(n+1)/2}$ on the diagonal. Let $\hat{C}$ be the matrix of $\hat{c}$ with respect to the basis $(\hat{\delta}_1,\dots,\hat{\delta}_m)$. Then the following proposition can be derived from [Bourbaki$_2$, Chap. V, §6, Exercice 3] (see also [Levine$_1$]):

<u>Propostion 1.6.3.</u> $\hat{C} = (-1)^{n+1} V^{-1} V^t$

<u>Corollary 1.6.4.</u> $\ker(\hat{c} - id_{\hat{H}}) = \ker \hat{H}$

<u>Proof.</u> If one denotes by $\mathbf{1}$ the $m \times m$ unit matrix, then it follows from Proposition 1.6.3:

$$\hat{C} - \mathbf{1} = -V^{-1}(V + (-1)^n V^t) \ .$$

From this, the corollary can be deduced immediately.

We denote by $c_{\mathbb{C}}$ respectively $\hat{c}_{\mathbb{C}}$ the complex monodromy operator $c_{\mathbb{C}} : H_{\mathbb{C}} \longrightarrow H_{\mathbb{C}}$, respectively the complex relative monodromy operator

$\hat{c}_{\mathbb{C}}:\hat{H}_{\mathbb{C}} \longrightarrow \hat{H}_{\mathbb{C}}$, where $H_{\mathbb{C}} := H \otimes \mathbb{C}$ and $\hat{H}_{\mathbb{C}} := \hat{H} \otimes \mathbb{C}$. By $[\hat{Le}_3]$,$[\hat{Le}_4]$ (see also [Looijenga$_3$, Theorem (5.14)]) the complex monodromy operator $c_{\mathbb{C}}$ is quasi-unipotent, i.e. its eigenvalues are roots of unity. By (1.6.2) therefore also the relative monodromy operator $\hat{c}_{\mathbb{C}}$ is quasi-unipotent.

If the singularity $(X,0)$ is given by quasi-homogeneous polynomials f_i, $i = 1,\dots,k$, of degree d_i with respect to the weights $wt(z_j) = w_j$, $j = 1,\dots,n+k$, for coordinates $(z_1,\dots,z_{n+k})$ of $\mathbb{C}^{n+k}$, and the singularity $(X',0)$ by the equations $f_i = 0$, $i = 1,\dots,k-1$, then we say briefly:

$(X,0)$ and $(X',0)$ are _quasi-homogeneous_ (with degrees $\underset{\sim}{d} = (d_1,\dots,d_k)$ and weights $\underset{\sim}{w} = (w_1,\dots,w_{n+k})$).

We denote by $\mu_0(X)$, respectively $\mu_0(X')$, the dimensions of the kernels (radicals) of the Milnor lattices of the corresponding singularities. Dimca has proven the following theorem [Dimca$_2$].

Theorem 1.6.5 (Dimca). _Let the singularities_ $(X,0)$ _and_ $(X',0)$ _be quasi-homogeneous of degrees_ $(d_1,\dots,d_k)$. _Then the following is true:_

(i) _The complex monodromy operator_ $c_{\mathbb{C}}$ _is diagonalizable, and its eigenvalues are_ d_k-th _roots of unity._

(ii) $\dim(\ker(c - id)) = \mu_0(X) + \mu_0(X')$.

Concerning the computation of Dynkin diagrams, we shall only be interested in the case that the singularity $(X',0)$ is an ordinary double point, hence a singularity of type A_1. In this case we show as an application of the above results:

Proposition 1.6.6. _Let the singularities_ $(X,0)$ _and_ $(X',0)$ _be quasi-homogeneous, and let_ $(X',0)$ _be a singularity of type_ A_1, _so_ $\mu' = \text{rk } H' = 1$. _Then the following is true:_

(i) _If the dimension_ n _is even, then_ $\hat{c}_{\mathbb{C}}$ _has exactly one non-trivial Jordan block. This is of the form_

$$\begin{pmatrix} 1 & 1 \\ 0 & 1 \end{pmatrix} .$$

_If_ $\hat{c}_{\mathbb{C}} = \hat{c}_s\hat{c}_u$ _is the (multiplicative) Jordan decomposition of_ $\hat{c}_{\mathbb{C}}$ _into a semi-simple part_ $\hat{c}_s$ _and a unipotent part_ $\hat{c}_u$, _then_

$$H' = \operatorname{im}(\hat{c}_u - id) \cap \hat{H} = \ker(\hat{c} - id) \cap \operatorname{im}(\hat{c} - id) .$$

(ii) <u>If the dimension</u> n <u>is odd, then also</u> $\hat{c}_{\mathbb{C}}$ <u>is diagonalizable.</u>

<u>Proof.</u> (i) If n is even, then the dimension of $(X',0)$ is odd. Therefore the intersection form on the Milnor fibre of $(X',0)$ is skew-symmetric and thus $\mu_0(X') = \mu' = 1$. Corollary 1.6.4 and Theorem 1.6.5 imply that

$$\dim(\ker(\hat{c} - \mathrm{id})) = \dim(\ker(c - \mathrm{id})).$$

From the exact sequence (1.6.2) one can deduce that

$$H' \subset \ker(\hat{c} - \mathrm{id}) \cap \mathrm{im}(\hat{c} - \mathrm{id}) \;,$$

$$\ker(\hat{c} - \mathrm{id})/H' \subset \ker(c - \mathrm{id}) \;,$$

$$\mathrm{im}(\hat{c} - \mathrm{id})/H' = \mathrm{im}(c - \mathrm{id}) \;.$$

Since $c_{\mathbb{C}}$ is diagonalizable, one has $\ker(c - \mathrm{id}) \cap \mathrm{im}(c - \mathrm{id}) = \{0\}$, and hence assertion (i).

(ii) If the dimension n is odd, then the dimension of $(X',0)$ is even. Thus $\mu_0(X') = 0$, and Corollary 1.6.4 and Theorem 1.6.5 imply that

$$\dim(\ker(\hat{c} - \mathrm{id})) = \dim(\ker(c - \mathrm{id})) + 1.$$

Then one can deduce from the exact sequence (1.6.2) that

$$\dim(\mathrm{im}(\hat{c} - \mathrm{id})) = \dim(\mathrm{im}(c - \mathrm{id})).$$

This yields assertion (ii) of Proposition 1.6.6.

Under the assumptions of this proposition, H' is determined by $\hat{c}$ for n even, for odd dimension n we don't obtain any additional informations except that H' is contained in the eigenspace of $\hat{c}$ corresponding to the eigenvalue 1 (1.6.2). Proposition 1.6.6 (i) should be compared with Proposition 2.4.3, a result due to K. Saito [Saito_1].

2. COMPUTATION OF DYNKIN DIAGRAMS

2.1. Polar curves

We consider in this section polar curves of isolated complete intersection singularities and invariants associated with these curves. This section serves for the preparation of the method for the computation of Dynkin diagrams described inthe next section.

Let $(X,0)$ be an isolated complete intersection singularity defined by a map-germ

$$f = (f_1,\ldots,f_k) : (\mathbb{C}^{n+k},0) \longrightarrow (\mathbb{C}^k,0) \quad .$$

We assume that also

$$f' := (f_1,\ldots,f_{k-1}) : (\mathbb{C}^{n+k},0) \longrightarrow (\mathbb{C}^{k-1},0)$$

defines an isolated complete intersection singularity denoted by $(X',0)$. Let $\zeta:\mathbb{C}^{n+k} \longrightarrow \mathbb{C}$ be a linear function, which can be chosen as the last coordinate function after a suitable change of coordinates. We consider the mapping

$$\Phi = (\Phi_1,\Phi_2) : X' \longrightarrow V \subset \mathbb{C}^2 \quad ,$$

where $\Phi_1 = \zeta$ and $\Phi_2 = f_k$. We denote by $\Sigma = \Sigma_\zeta(f_k)$ the critical locus of this mapping, and by $\Delta = \Delta_\zeta(f_k)$ the discriminant of this mapping. We assume furthermore that the linear function ζ is chosen in such a way that the following condition is satisfied:

(2.1.1) Φ is a submersion in each point of $\Phi^{-1}(0) - \{0\}$.

That this condition holds for a generic function ζ follows e.g. from [Looijenga$_3$, Lemma (5.2)]. Finally we assume that

(2.1.2) $f_k|_{X'}$ is not a submersion in 0.

Then one has the following facts (see e.g. [Looijenga$_3$, Theorem (2.8)]):

(i) Σ is a (not necessarily reduced) curve, i.e. $\dim \Sigma = 1$.

(ii) $\Delta = \Phi(\Sigma)$ is a plane curve

(iii) The mapping $\Phi|_\Sigma : \Sigma \longrightarrow \Delta$ is a finite mapping.

<u>Definition 2.1.3.</u> The curve $\Sigma_\zeta(f_k)$ is called the (relative) <u>polar curve</u> of f_k with respect to ζ. Its image $\Delta_\zeta(f_k)$ under Φ is called the (relative) <u>Cerf diagram</u> of f_k with respect to ζ.

Let $\Sigma = \cup \Sigma_i$ be the decomposition of the polar curve into irreducible components, and set $\Delta_i := \Phi(\Sigma_i)$. We choose coordinates of $\mathbb{C}^2$ denoted by (ζ,λ) such that the mapping Φ is given by $\Phi_1(z) = \zeta$ and $\Phi_2(z) = f_k(z) = \lambda$ for $z \in X'$.

We assume that the branch Δ_i does not coincide with the coordinate axis $\zeta = 0$. Then Δ_i has a Puiseux parametrization

$$\lambda = a_i \zeta^{\rho_i} + \dots ,$$

where $a_i \in \mathbb{C}$, $a_i \neq 0$, and $\rho_i \in \mathbb{Q}$. By definition of λ this can be interpreted in such a way that $f_k|_{\Sigma_i}$ can be written as a fractional power series in ζ beginning with the term $a_i\zeta^{\rho_i}$. If X' is non-singular, then all the branches Δ_i are tangent to the coordinate axis $\lambda = 0$ at the origin [Lê$_1$, Proposition 1.2], and therefore $\rho_i > 1$. But this is in general no longer true if X' is singular (cf. [Lê$_3$]).

If $\Delta_i = \{\zeta = 0\}$ then we set $\rho_i = 0$.

The method for the computation of Dynkin diagrams presented in the next section is based on the fact that the critical points of the following functions all lie on the polar curve $\Sigma = \Sigma_\zeta(f_k)$:

$$f_k - 2\varepsilon\zeta|_{X'} , \quad f_k|_{X' \cap \{\zeta = \varepsilon\}} , \quad f_k + (\zeta - \varepsilon)^2|_{X'} .$$

Here $0 \leq \varepsilon << 1$. Let μ_i (respectively ν_i) be the sum of the Milnor numbers of the critical points of $f_k - 2\varepsilon\zeta|_{X'}$ (respectively $f_k|_{X' \cap \{\zeta = \varepsilon\}}$), $\varepsilon \neq 0$, which belong to $\Sigma_i - \{0\}$ and tend to 0 for $\varepsilon \longrightarrow 0$. Then one has the following relation:

<u>Proposition 2.1.4.</u> <u>If</u> $\rho_i > 1$ <u>then</u> $\mu_i = \nu_i(\rho_i - 1)$.

<u>Proof.</u> Let Σ_i be an irreducible component of Σ with $\rho_i > 1$. Let n_i be the multiplicity of Σ_i. Since $\Phi|_\Sigma$ is finite, we may assume, after possibly shrinking X' and V, that $\Phi_{|\Sigma_i - \{0\}} : \Sigma_i - \{0\} \to \Delta_i - \{0\}$

is a covering of degree d_i. A critical point ξ of $f_k - 2\varepsilon\zeta|_{X'}$ lying on $\Sigma_i - \{0\}$ is mapped under Φ to a point on $\Delta_i - \{0\}$, such that the tangent to Δ_i in this point is parallel to the line $\lambda - 2\varepsilon\zeta = 0$. Let m_i be the number of such points on $\Delta_i - \{0\}$. By [Teissier$_1$, 2.6] the Milnor number of the critical point ξ is equal to n_i. Therefore one has

$$\mu_i = m_i d_i n_i \ .$$

The critical points of $f_k|_{X' \cap \{\zeta = \varepsilon\}}$ which lie on $\Sigma_i - \{0\}$ are the points of $\Sigma_i \cap \{\zeta = \varepsilon\}$. They are mapped under Φ to the points of $\Delta_i \cap \{\zeta = \varepsilon\}$. The Milnor number of such a critical point is equal to n_i for the same reason as above. Therefore one has

$$\nu_i = d_i n_i (\Delta_{i,red} \cdot \{\zeta = \varepsilon\}) = d_i (\Delta_i \cdot \{\zeta = 0\}),$$

where $(\cdot)$ denotes the intersection number and $\Delta_{i,red}$ the reduced curve corresponding to Δ_i. Since Δ_i is a branch of a plane curve and $\rho_i > 1$, one can easily verify that

$$n_i m_i = (\Delta_i \cdot \{\lambda = 0\}) - (\Delta_i \cdot \{\zeta = 0\}) \ .$$

Therefore one has

$$\rho_i = \frac{(\Delta_i \cdot \{\lambda = 0\})}{(\Delta_i \cdot \{\zeta = 0\})} = \frac{n_i m_i + (\Delta_i \cdot \{\zeta = 0\})}{(\Delta_i \cdot \{\zeta = 0\})}$$

$$= \frac{d_i n_i m_i + d_i (\Delta_i \cdot \{\zeta = 0\})}{d_i (\Delta_i \cdot \{\zeta = 0\})} = \frac{\mu_i + \nu_i}{\nu_i} \ ,$$

which had to be proved.

If $g : (\mathbb{C}^{n+k},0) \longrightarrow (\mathbb{C}^k,0)$ defines an isolated complete intersection singularity $(Y,0)$, then we denote the Milnor number of $(Y,0)$ by $\mu(g)$ as well. We denote by X'_t a Milnor fibre of $(X',0)$ with respect to the mapping f', where $t \in \mathbb{C}^{k-1}$ is sufficiently small and not contained in the discriminant $D_{f'}$ of f'. By [Lê$_2$] (see also [Dimca$_1$, Proposition 1.1]) the number m of critical points of the function

$$f_k - 2\varepsilon\zeta|_{X'_t}$$

(where one has to count the points with their Milnor numbers) is equal to

$$m = \mu(f) + \mu(f').$$

<u>Proposition 2.1.5.</u> <u>One has the following relations:</u>

(i) $\mu(f) = \mu(f', f_k - 2\varepsilon\zeta) + \sum_{i:\rho_i > 1} \mu_i .$

(ii) <u>If</u> $a_i \neq -1$ <u>for</u> $\rho_i = 2$, <u>then</u>

$$\mu(f', f_k + \zeta^2) = \mu(f', f_k - 2\varepsilon\zeta) + \sum_{i:1<\rho_i<2} \mu_i + \sum_{i:\rho_i \geq 2} \nu_i .$$

<u>Proof.</u> (i) The number of critical points of the function $f_k - 2\varepsilon\zeta|_{X_t'}$ tending to 0 for $t \longrightarrow 0$ is equal to the number

$$\mu(f', f_k - 2\varepsilon\zeta) + \mu(f')$$

by the same result as above. Note that for a branch Σ_i with $\rho_i \leq 1$ the number m_i, which was defined in the proof of Proposition 2.1.4, and therefore also the number μ_i, is equal to zero. Therefore the Milnor numbers of the critical points of $f_k - 2\varepsilon\zeta|_{X_t'}$ which do not tend to 0 for $t \longrightarrow 0$ add up to

$$\sum_{i:\rho_i>1} \mu_i .$$

This yields the formula of (i).

(ii) The formula of (ii) follows from (i), Proposition 2.1.4, and the equality

$$\mu(f', f_k + \zeta^2 - 2\varepsilon\zeta) = \mu(f', f_k - 2\varepsilon\zeta).$$

2.2. A generalization of a method of Gabrielov

In [Gabrielov$_3$] A.M. Gabrielov presents a method for the computation of a Dynkin diagram of an isolated hypersurface singularity given by a function-germ $f:(\mathbb{C}^{n+1},0) \longrightarrow (\mathbb{C},0)$ from a Dynkin diagram of the singularity defined by $f + \zeta^2$. We generalize his method to the above situation: A germ of a function $f_k:(X',0) \longrightarrow (\mathbb{C},0)$ with an isolated singularity at 0 defined on an isolated complete intersection singularity $(X',0)$.

Gabrielov's results and proofs carry over to the more general situation with only minor modifications. Therefore we mainly follow his article. For the statement of the results we first consider special Dynkin diagrams of the singularity defined by $(f',f_k+\zeta^2)$.

Proposition 2.2.1. We consider the small perturbation $g_\varepsilon = f_k + (\zeta-\varepsilon)^2|_{X'}$, $0 \leq \varepsilon << 1$, of $f_k + \zeta^2|_{X'}$. The critical values of the critical points of g_ε lying on $\Sigma_i - \{0\}$ and tending to 0 for $\varepsilon \longrightarrow 0$ are

$$\text{if} \quad \rho_i > 2 : a_i\varepsilon^{\rho_i} + o(\varepsilon^{\rho_i}) \, ,$$

$$\text{if} \quad \rho_i = 2 \quad (\text{we assume} \quad a_i \neq -1 \quad \text{in this case}):$$

$$\varepsilon^2 a_i/(a_i+1) + o(\varepsilon^2) \, ,$$

$$\text{if} \quad \rho_i < 2 : \varepsilon^2 + o(\varepsilon^2) \, .$$

The critical point 0 is mapped to ε^2 by g_ε.

Proof. The critical points of g_ε lying on $\Sigma_i - \{0\}$ are the solutions of the equation

$$\frac{\partial g_\varepsilon|\Sigma_i}{\partial\zeta} = a_i\rho_i\zeta^{\rho_i-1} + 2(\zeta-\varepsilon) + o(\zeta^{\rho_i-1}) = 0,$$

which tend to 0 for $\varepsilon \longrightarrow 0$. Considering the asymptotics of these solutions for $\varepsilon \longrightarrow 0$ and their images under g_ε, one obtains the above values.

By Proposition 2.2.1 the critical values of g_ε are contained in annuli around the origin: Let R be the set of all the values ρ_i. Let $\varepsilon \neq 0$ be sufficiently small. For each $\rho \in R$ with $\rho > 2$ we can choose positive numbers q'_ρ and q''_ρ, such that the critical values of all the critical points of g_ε belonging to $\Sigma_i - \{0\}$ with $\rho_i = \rho$ and tending to 0 for $\varepsilon \longrightarrow 0$ are contained in the open annulus $\{u \mid q'_\rho < |u| < q''_\rho\}$. We can also choose positive numbers q'_2 and q''_2 such that the critical values of all the critical points belonging to $\Sigma_i - \{0\}$ with $\rho_i \leq 2$ and tending to 0 for $\varepsilon \longrightarrow 0$, and of the critical point 0, are contained in the open annulus $\{u \mid q'_2 < |u| < q''_2\}$. Moreover, the number q'_ρ and q''_ρ $(\rho \geq 2)$ can be chosen in such a way that $q''_\rho < q'_\rho$ holds for $\rho > \tilde{\rho}$.

In what follows, the argument of the complex number u, $\arg u$, is supposed to lie in the interval $[-\pi,\pi]$. Let $\sigma: \mathbb{R}_+ \longrightarrow \mathbb{R}_+$ be a continuous monotonously non-increasing function with $\sigma(q) = \rho - 1$ for $q'_\rho \le q \le q''_\rho$. We define

$$V_r := \{u \mid \arg u + 2\pi\sigma(|u|) \geq (2r-1)\pi\}$$

for $r = 0,1,2,\ldots$.

We assume that the following condition is satisfied:

(2.2.2) $\quad (-a_i)^{\ell_i} \notin \mathbb{R}_+ \quad$ for $\rho_i = k_i/\ell_i$, $(k_i,\ell_i) = 1$.

Then one can easily deduce from Proposition 2.2.1 that for a sufficiently small $\varepsilon \in \mathbb{R}_+ - \{0\}$ the critical values of g_ε are not contained in $\mathbb{R}_-$ or in the boundary of a set V_r.

Now we consider the function

$$g_{\varepsilon,t} = f_k + (\zeta - \varepsilon)^2 |_{X'_t} \quad .$$

Choose $\varepsilon >> \delta > 0$ small enough, such that for all $t \in \mathbb{C}^{k-1}$ with $0 < |t| \le \delta$ and $t \notin D_{f'}$ (where $D_{f'}$ denotes the discriminant of f') the critical values of $g_{\varepsilon,t}$ do not lie in $\mathbb{R}_-$, on the boundary of an annulus, or on the boundary of a set V_r . Choose such a t. If $g_{\varepsilon,t}$ is not a Morse function, i.e. if it does not have only non-degenerate critical points with distinct critical values, then we replace this function by a nearby Morse function (this is possible by [Dimca$_1$]).

Now we choose a strongly distinguished system of paths $(\varphi_1,\ldots,\varphi_\nu)$ joining the critical values of $g_{\varepsilon,t}$ to the non-critical value 0. Strongly distinguished means in this case that the paths are non-self-intersecting, have no mutual intersection points except 0 ($\varphi_i \cap \varphi_j = \{0\}$), and are ordered in the following way: Let $\arg u_i > \arg u_j$ for $i < j$, where u_i, respectively u_j, is the intersection point of φ_i , respectively φ_j, with a small circle centered at the origin. We assume in addition that the system of paths has the following property:

(V) Each path intersects $\mathbb{R}_-$ only at the origin, and a path starting at a critical value contained in V_r stays entirely in V_r.

This condition is e.g. satisfied by a system of line segments between the critical values of $g_{\varepsilon,t}$ and 0 (with suitable modifications if two critical values lie on the same line segment). The whole situation is illustrated for an example in Figure 2.2.1.

$$f(x,y,z)=(2xz+y^2, x^2-b^2y^2z+\frac{1}{\sqrt{2}}byz^4+b^4z^4),\ b^8=-1\ (K^{\#}_{1,3})$$

$$\rho_1 = 4,\ \rho_2 = \frac{11}{2},\ \nu_1 = 1,\ \nu_2 = 2,$$

$$\mu(f_1, f_2-2\varepsilon z) = 2$$

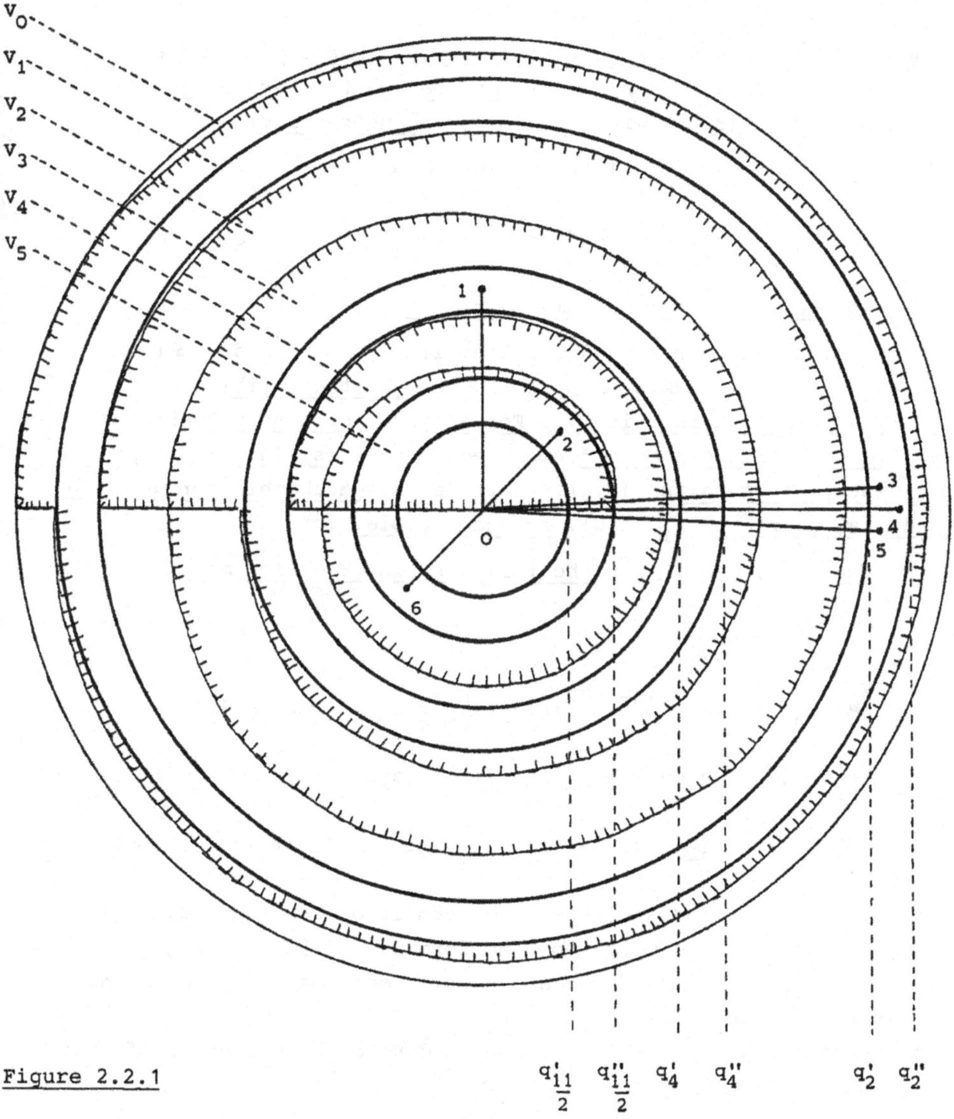

Figure 2.2.1

After these preparations we can formulate the following theorem.

Theorem 2.2.3. *Let* $f = (f_1,\dots,f_k):(\mathbb{C}^{n+k},0) \longrightarrow (\mathbb{C}^k,0)$ *be a map-germ defining an isolated complete intersection singularity* $(X,0)$, *and let* $f' = (f_1,\dots,f_{k-1})$ *define an isolated complete intersection singularity* $(X',0)$. *Let* $\zeta : \mathbb{C}^{n+k} \longrightarrow \mathbb{C}$ *be a linear function, and assume that the conditions* (2.1.1), (2.1.2), *and* (2.2.2) *are satisfied. Finally, let* $(e_1,\dots,e_\nu)$ *be a basis of thimbles for* $g = (f', f_k + \zeta^2)$ *corresponding to a strongly distinguished system of paths satisfying condition* (V).

Then there exists a strongly distinguished basis of thimbles $(e_j^r \mid 1 \leq j \leq \nu,\ 1 \leq r \leq M_j)$ *for* $(X,0)$, *ordered by the lexicographic order of the pairs* (r,j), *with the following properties:*

(a) *For* $1 \leq j, j' \leq \nu$, $1 \leq r \leq M_j$ *one has*

$$\langle e_j^1, e_{j'}^1\rangle = \langle e_j, e_{j'}\rangle \quad \underline{\text{and}} \quad e_j^{r+1} = \hat{c}(e_j^r) \ ,$$

where $\hat{c}$ *is the relative monodromy operator of* (X,X').

(b) *Call a pair* (r,j) *admissible, if* $1 \leq j \leq \nu$ *and* $1 \leq r \leq M_j$. *Then a pair* (r,j) *is admissible if and only if the thimble* e_j *is defined by a path contained in* V_r. *For a system of line segments as above, this condition can be reformulated as follows: The first* μ_i *pairs of each set of pairs* (r,j), *where* e_j *is a thimble corresponding to a critical point on* $\Sigma_i - \{0\}$, *are admissible.*

(c) *The basis* (e_j^r) *has the following intersection matrix:*

$$\langle e_j^r, e_{j'}^r\rangle = \langle e_j, e_{j'}\rangle \ ,$$

$$\langle e_j^r, e_j^{r'}\rangle = -(-1)^{n(n-1)/2}(r'-r)^n \quad \underline{\text{for}} \quad |r'-r| = 1,$$

$$\langle e_j^r, e_{j'}^{r'}\rangle = -\langle e_j, e_{j'}\rangle \ \underline{\text{for}}\ |r'-r| = 1 \quad \underline{\text{and}} \quad (r'-r)(j'-j) < 0,$$

$$\langle e_j^r, e_{j'}^{r'}\rangle = \quad \underline{\text{for}} \quad |r'-r| > 1 \quad \underline{\text{or}} \quad (r'-r)(j'-j) > 0.$$

Note that the second and the third formula of Theorem 2.2.3 (c) differ by a sign from the corresponding formulas of [Gabrielov$_3$, Theorem 1]. The signs given there are not correct, as is shown in the proof of the following lemma.

The remaining part of this section is devoted to the proof of this theorem. Assertion (c) follows easily from assertion (a) as indicated in the following lemma.

Lemma 2.2.4. (cf. [Gabrielov$_3$, Lemma 1]). Let $(e_j^r \mid 1 \le j \le \nu,\ 1 \le r \le M_j)$ be a strongly distinguished basis of thimbles, ordered by the lexicographic order of the pairs (r,j), with the property: $e_j^{r+1} = \hat{c}(e_j^r)$ for $1 \le r \le M_j - 1$, where $\hat{c}$ is the corresponding Coxeter element. Then

$$\langle e_j^r, e_{j'}^r \rangle = \langle e_j^1, e_{j'}^1 \rangle ,$$

$$\langle e_j^r, e_j^{r'} \rangle = -(-1)^{n(n-1)/2}(r'-r)^n \quad \text{for} \quad |r'-r| = 1 ,$$

$$\langle e_j^r, e_{j'}^{r'} \rangle = -\langle e_j^1, e_{j'}^1 \rangle \quad \text{for} \quad |r'-r| = 1 \quad \text{and} \quad (r'-r)(j'-j) < 0,$$

$$\langle e_j^r, e_{j'}^{r'} \rangle = 0 \quad \text{for} \quad |r'-r| > 1 \quad \text{or} \quad (r'-r)(j'-j) > 0 .$$

Proof. Since $\hat{c}$ preserves the intersection form, one has $\langle e_j^r, e_{j'}^r \rangle = \langle e_j^1, e_{j'}^1 \rangle$. Setting

$$\hat{\gamma}_j^r = s^{(n)}_{e_j^r} ,$$

one has $\hat{c} = \hat{\gamma}_1^1 \hat{\gamma}_2^1 \ldots \hat{\gamma}_\nu^{M_\nu}$. Then $\hat{c}(e_j^r) = e_j^{r+1}$ implies that

$$\hat{\gamma}_{j'}^{r'}(e_j^r) = e_j^r \quad \text{for} \quad (r',j') > (r+1,j) ,$$

$$\hat{\gamma}_j^{r+1}(e_j^r) = e_j^r - (-1)^{n(n-1)/2} \langle e_j^r, e_j^{r+1} \rangle e_j^{r+1} = e_j^r + e_j^{r+1} ,$$

$$\hat{\gamma}_{j'}^{r'}(e_j^r + e_j^{r+1}) = e_j^r + e_j^{r+1} \quad \text{for} \quad r' = r+1,\ j' < j \quad \text{and}$$
$$r' = r,\ j' > j .$$

The other formulas of Lemma 2.2.4 can be derived from these equations.

Therefore, for the proof of Theorem 2.2.3, one has to construct a basis of thimbles (e_j^r) for f with the properties (a) and (b). For this purpose we consider the perturbation $f_\varepsilon = f_k - 2\varepsilon\zeta|_{X'}$ of the function $f_k|_{X'}$. Again we compute the asymptotics of the critical values of this function for $\varepsilon \longrightarrow 0$.

Proposition 2.2.5. The critical values of the critical points of $f_\varepsilon = f_k - 2\varepsilon\zeta|_{X'}$ lying on $\Sigma_i - \{0\}$ and tending to 0 for $\varepsilon \longrightarrow 0$

are

$$-a_i(\rho_i - 1)\left(\frac{2\varepsilon}{a_i\rho_i}\right)^{\rho_i/(\rho_i-1)} + o(\varepsilon^{\rho_i/(\rho_i-1)}).$$

Here $\rho_i > 1$ is a necessary condition for the existence of such points. The critical point 0 is mapped to 0 by f_ε .

Proof. This follows from an analogous calculation as for the proof of Proposition 2.2.1.

As above, this proposition implies that we can find for a small $\varepsilon \neq 0$ positive numbers Q'_ρ and Q''_ρ for $\rho \in R$, $\rho > 2$, and a positive number Q_2 such that $Q''_\rho > Q'_{\tilde{\rho}}$ for $\rho > \tilde{\rho}$, and such that the following conditions are satisfied: The critical values of all the critical points on f_ε lying on $\Sigma_i - \{0\}$ with $\rho_i = \rho > 2$ (and tending to 0 for $\varepsilon \longrightarrow 0$) are contained in an open annulus $\{u \mid Q''_\rho < |u| < Q'_\rho\}$. The critical value 0 and the critical values of the critical points of f_ε lying on $\Sigma_i - \{0\}$ with $\rho_i \leq 2$ (and tending to 0 for $\varepsilon \longrightarrow 0$) are contained in an open disc $\{u \mid |u| < Q'_2\}$. Note that the order between the annuli given by $\rho \in R$ is reversed compared to the situation for g_ε , and 0 has become a critical value.

The basic idea is now to relate the two perturbations g_ε and f_ε by the deformation

$$F^\eta_\varepsilon = f_k + \eta\zeta^2 - 2\varepsilon\zeta|_{X'} \quad , \eta \in [0,1] \quad ,$$

where $F^0_\varepsilon = f_\varepsilon$, $F^1_\varepsilon = g_\varepsilon - \varepsilon^2$. Now it turns out that there is a domain G in the image space of f_ε such that the critical values of f_ε contained in G and only they pass to the critical values of $g_\varepsilon - \varepsilon^2$ tending to 0 for $\varepsilon \longrightarrow 0$, when η varies from 0 to 1. We shall define G and consider systems of paths in G related to the systems of paths defined previously for g_ε .

We choose a base point $u^* \in \mathbb{R}_-$, which is a non-critical value of f_ε with $u^* < -\max_\rho Q'_\rho$. Let $\xi : [0,-u^*] \longrightarrow \mathbb{R}_+$ be a continuous monotonously non-increasing function with

$$\begin{aligned}
\xi(q) &= 1 && \text{for } q < Q'_2 \quad , \\
\xi(q) &= 1/(\rho - 1) && \text{for } Q''_\rho \leq q \leq Q'_\rho \quad , \qquad \rho > 2, \\
\xi(q) &> 0 && \text{for } \rho < -u^* \quad , \\
\xi(-u^*) &= 0 \quad .
\end{aligned}$$

Set $G = \{u \mid |u| \leq -u^*,\ |\arg(-u)| \leq \xi(|u|)\pi\}$. Moreover, set $W_r = \{u \in G \mid \arg(-u) + (2r-1)\xi(|u|)\pi \leq 2\pi\}$ for $r = 1,2,\ldots$. In particular $W_1 = G$.

For later purposes we also need a family of homeomorphisms $\tau_\varphi : \mathbb{C} \longrightarrow \mathbb{C}$ defined by

$$\tau_\varphi(u) = u \exp(2\pi\sqrt{-1}\,\varphi\,(\xi(|u|) + 1)).$$

By assumption the condition (2.2.2) is satisfied. Let ε be a small positive number. Then one can easily deduce from Proposition 2.2.5:

<u>Lemma 2.2.6.</u> <u>No critical values of</u> f_ε <u>are contained in the boundaries of the sets</u> $\tau_\varphi(W_r)$ <u>for</u> $\varphi = 0,1,\ldots$ <u>and arbitrary</u> $r \in \mathbb{N}$.

The same is true for the function

$$f_{\varepsilon,t} = f_k - 2\varepsilon\zeta|_{X_t'}$$

for $0 < |t| \leq \delta$ (after possibly shrinking δ) or for a nearby Morse function. We also assume that δ is chosen small enough such that no critical values of these functions lie on the boundaries of the annuli or on the boundary of the disc of radius ϱ_2'.

We choose a strongly distinguished system of paths $(\tilde{\varphi}_1,\ldots,\tilde{\varphi}_{\tilde{\nu}})$ inside G connecting the critical values of $f_{\varepsilon,t}$ contained in G with the non-critical value u^*. Strongly distinguished means in this case that the paths are non-self-intersecting, any two meet only in u^*, and the paths are ordered as follows: For $i < j$ one has $\arg(u_i - u^*) > \arg(u_j - u^*)$, where u_i, respectively u_j, is the intersection point of $\tilde{\varphi}_i$, respectively $\tilde{\varphi}_j$, with a small circle centered at u^*. We assume in addition that the following condition is satisfied

(W) a path starting at a critical value belonging to W_r stays entirely in W_r.

Let (e_j^1) be a system of thimbles of f defined by the chosen system of paths. Then we have the following lemma.

<u>Lemma 2.2.7</u> (cf. [Gabrielov$_3$, Lemma 2]). <u>Let the parameter</u> η <u>vary from</u> 0 <u>to</u> 1 <u>in the family</u>

$$F^\eta_{\varepsilon,t} = f_k + \eta\zeta^2 - 2\varepsilon\zeta|_{X_t'} \quad .$$

(i) Then exactly the critical values of $F^0_{\varepsilon,t} = f_{\varepsilon,t}$ contained in G and tending to 0 for $\varepsilon \longrightarrow 0$ and $t \longrightarrow 0$ pass to the critical values of $F^1_{\varepsilon,t} = g_{\varepsilon,t} - \varepsilon^2$ tending to 0 for $\varepsilon \longrightarrow 0$ and $t \longrightarrow 0$. Exactly those lying in W_r pass to those contained in $V_r - \varepsilon^2 = \{u - \varepsilon^2 | u \in V_r\}$.

(ii) A strongly distinguished system of paths for $F^0_{\varepsilon,t}$ satisfying the condition (W) can be homotopically deformed into a system of paths for $g_{\varepsilon,t} - \varepsilon^2$ which can be obtained by a translation by $(-\varepsilon^2)$ from a strongly distinguished system of paths for $g_{\varepsilon,t}$ satisfying the condition (V).

(iii) The systems of thimbles (e^1_j) and (e_j) defined by the corresponding systems of paths have the same intersection matrices.

Since $|t| \leq \delta << \varepsilon$, the assertions of the lemma follow from the corresponding assertions for $t = 0$. The Milnor numbers of the critical point 0 of g_ε and f_ε are the same by Proposition 2.1.5. Therefore this lemma is implied by [Gabrielov$_3$, Lemma 2].

Now we consider the action of the relative monodromy. Let $X \subset \mathbb{C}^{n+k+h}$, $S \subset \mathbb{C}^{h+k}$, and let

$$F : X \longrightarrow S$$

be a representative of the semi-universal deformation of f in normal form, i.e. for suitable coordinates $x = (x_1,\dots,x_{n+k})$ and $\lambda = (\lambda_1,\dots,\lambda_h)$ one has:

$$F_j(x,\lambda) = \begin{cases} \lambda_j & \text{for } j = 1,\dots,h \ , \\ f_{j-h}(x) + \sum_{i=1}^{h} g_{i,j-h}(x)\lambda_i & \text{for } j = h+1,\dots,h+k, \end{cases}$$

where $g_{i,j-h} \in m\, \mathcal{O}_{\mathbb{C}^{n+k},0}$ are suitable functions with

$$(g_{11},\dots,g_{1k}) = (0,\dots,0,\zeta).$$

We consider the geometric monodromy corresponding to the closed path

$$\omega : S^1 \longrightarrow S - D$$

$$\varphi \longrightarrow (\lambda,y) = (\lambda_1,\dots,\lambda_h;y_1,\dots,y_k)$$

with $\lambda_1 = 2\varepsilon e^{2\pi\sqrt{-1}\varphi}$, $\lambda_j = 0$ for $1 < j \leq h$, $y_j = 0$ for $1 \leq j < k$, and $y_k = u^* e^{2\pi\sqrt{-1}\varphi}$. Since this loop is homotopic to the corresponding loop with $\lambda_1 = 0$ in $S - D$, it also defines the relative monodromy operator $\hat{c}$.

We consider the disc $\mathbb{D} = \{y_k \mid |y_k| \leq -u^*\}$ and the point

$$t_\varphi = (\lambda_1, \ldots, \lambda_h; y_1, \ldots, y_{k-1}) \text{ with } \lambda_1 = 2\varepsilon e^{2\pi\sqrt{-1}\varphi}, \lambda_2 = \ldots = \lambda_h = 0,$$

$$(y_1, \ldots, y_{k-1}) = t.$$

To the closed path ω corresponds a continuous family of homeomorphisms

$$\Theta_\varphi : \mathbb{D} \longrightarrow \mathbb{D}$$

with the properties that $\Theta_0 = \mathrm{id}$, Θ_φ maps the points $u \in \mathbb{D}$ with $(t_0, u) \in D$ (critical values of $f_{\varepsilon,t}$) to the corresponding point $\Theta_\varphi(u)$ with $(t_\varphi, \Theta_\varphi(u)) \in D$ (critical values of $f_{\varepsilon\exp(2\pi\sqrt{-1}\varphi),t}$), and $\Theta_\varphi(u^*) = u^* \exp(2\pi\sqrt{-1}\varphi)$. Moreover, we can choose Θ_φ in such a way that Θ_φ coincides with τ_φ on the boundaries of the sets W_r.

Lemma 2.2.8. The sets $\tau_{r-1}(W_r)$ and $\tau_{r'-1}(W_{r'})$ have no common inner points for $r \neq r'$. Each critical value of $f_{\varepsilon,t}$ is contained in a set $\tau_{r-1}(W_r)$ for some r.

Proof. One can easily show that the interior of the set $\tau_{r-1}(W_r)$ coincides with the interior of the set

$$\tau_{r-1}(G) - \Big(\bigcup_{0 \leq p \leq r-2} \tau_p(G) \Big) .$$

Each critical value of $f_{\varepsilon,t}$ is contained in a set $\tau_{r-1}(G)$ for some r, but not in the boundary of a set $\tau_p(G)$. This proves the lemma.

We extend the system of paths in G defined above to a system of paths between the critical values of $f_{\varepsilon,t}$ and u^* as follows: For a critical value u of $f_{\varepsilon,t}$ contained in $\tau_{r-1}(W_r) = \Theta_{r-1}(W_r)$, we define the path between this value and u^* as the result of applying Θ_{r-1} to the path between $\Theta_{r-1}^{-1}(u)$ and u^* contained in W_r. The order between these paths is defined as above. Using Lemma 2.2.8, one can easily check that this defines a strongly distinguished system of paths for the singularity $(X,0)$ (cf. Figure 2.2.2).

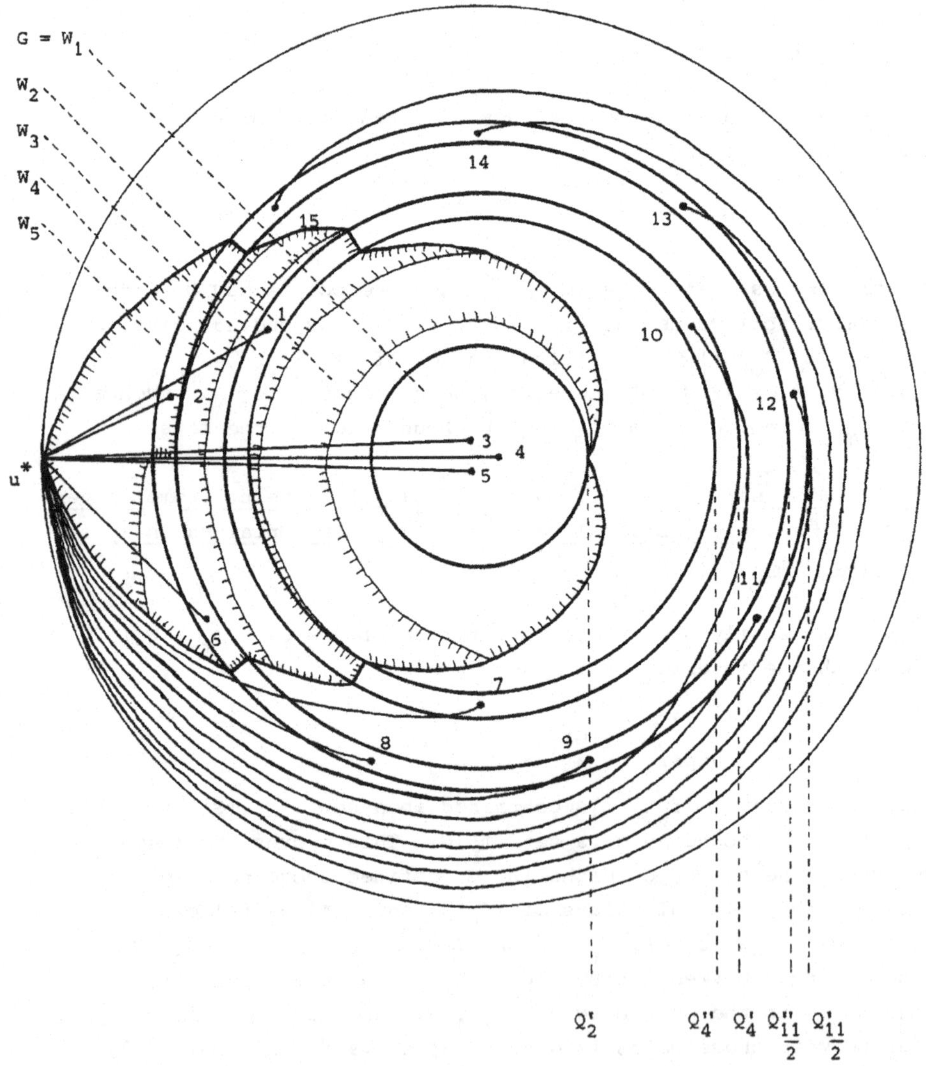

Figure 2.2.2 (cf. Figure 2.2.1)

Let (e_j^r) be a corresponding strongly distinguished basis of thimbles. It follows from the above construction that this basis is ordered by the lexicographic order of the pairs (r,j) and has the properties of the assertions (a) and (b) of Theorem 2.2.3.

This completes the proof of Theorem 2.2.3.

2.3. Dynkin diagrams of an intersection of two quadrics

By Theorem 2.2.3 the computation of Dynkin diagrams can be reduced to the computation of Dynkin diagrams of simpler singularities. The in some sense simplest non-hypersurface complete intersection singularities are the isolated singularities of an intersection of two quadrics. For these occur in the semi-universal deformation of any isolated complete intersection singularity, which is not a hypersurface singularity. These are the singularities $(X,0)$ given by a mapping

$$f = (f_1, f_2) : \mathbb{C}^{n+2} \longrightarrow \mathbb{C}^2$$

with

$$f_1(z) = z_1^2 + z_2^2 + \ldots + z_{n+2}^2 ,$$

$$f_2(z) = a_1 z_1^2 + a_2 z_2^2 + \ldots + a_{n+2} z_{n+2}^2 ,$$

where $a_i \in \mathbb{C}$, $a_i \neq a_j$ for $i \neq j$, $1 \leq i,j \leq n+2$. Their deformation theory has been studied by H. Knörrer [Knörrer$_1$] (and recently also by Y. Mérindol [Mérindol$_1$]). In this section we compute Dynkin diagrams of these singularities.

These singularities belong to the class of Brieskorn-Pham singularities, for which H. Hamm [Hamm$_2$] has given a basis of the complexified Milnor lattice $H_{\mathbb{C}} = H \otimes \mathbb{C}$, generalizing a method of F. Pham [Pham$_1$] in the hypersurface case. We briefly present Hamm's construction in this special case. For this purpose we assume that $a_j \in \mathbb{R}$ for all $1 \leq j \leq n+2$ and that $a_1 < a_2 < \ldots < a_{n+2}$.

We set $X' = f_1^{-1}(0)$. The function $f_2 : X' \longrightarrow \mathbb{C}$ is a submersion outside the origin. We denote the fibre of f_2 over a sufficiently small $\eta \in \mathbb{R}$, $\eta \neq 0$, with $Y = Y_\eta$. Hamm [Hamm$_2$, Lemma 2.4] shows that the Milnor fibre $X_{s_0} = Y \cap B_\varepsilon$ of $(X,0)$ for $s_0 = (0,\eta)$ is a deformation retract of Y.

Let $\Psi : \mathbb{C}^{n+2} \longrightarrow \mathbb{C}^{n+2}$ be the mapping defined by

$(z_1,\dots,z_{n+2}) \longrightarrow (w_1,\dots,w_{n+2}) = (z_1^2,\dots,z_{n+2}^2)$. Then Ψ maps X and X' to linear subspaces and Y to an affine subspace of $\mathbb{C}^{n+2}$. Let

$$Z' := \Psi(Y) \cap \mathbb{R}^{n+2} .$$

Then $\Psi^{-1}(Z')$ is a deformation retract of Y [Hamm$_2$, Lemma 3.1]. Let $W^0 = \mathbb{C}^{n+2}$ and $W^j = \{w \in \mathbb{C}^{n+2} \mid$ there are $r_1,\dots,r_j$ with $1 \leq r_1 < \dots < r_j \leq n+2$ and $w_{r_1} = \dots = w_{r_j} = 0\}$. We define a cellular decomposition of Z' as follows: The set $Z' \cap (W^{n-q} - W^{n-q+1})$ is a disjoint union of topological q-cells for $q = 0,1,\dots,n$; these form the cells of the cellular decomposition. This decomposition induces a cellular decomposition of $\Psi^{-1}(Z')$ as follows: Let $\tilde{\Psi}: \mathbb{R}^{n+2} \longrightarrow \mathbb{C}^{n+2}$ be defined by $(w_1,\dots,w_{n+2}) \longrightarrow (z_1,\dots,z_{n+2})$ with $z_i^2 = w_i$ and $z_i \geq 0$ or $-\sqrt{-1}\, z_i > 0$ for each i with $1 \leq i \leq n+2$. Obviously $\tilde{\Psi}$ is a homeomorphism onto its image. We use $\tilde{\Psi}$ to obtain a cellular decomposition of $\tilde{\Psi}(Z')$. Let

$$\tau_j : \mathbb{C}^{n+2} \longrightarrow \mathbb{C}^{n+2}$$

for $j = 1,\dots,n+2$ be the linear mapping which maps $(z_1,\dots,z_{n+2})$ to $(z_1,\dots,z_{j-1},-z_j,z_{j+1},\dots,z_{n+2})$, and let Ω be the subgroup of $GL(n+2,\mathbb{C})$ generated by $\tau_1,\dots,\tau_{n+2}$. Then $\Psi^{-1}(\mathbb{R}^{n+2}) = \Omega \cdot \tilde{\Psi}(\mathbb{R}^{n+2})$ and $\Psi^{-1}(Z') = \Omega \cdot \tilde{\Psi}(Z')$. Therefore we also get a cellular decomposition of $\Psi^{-1}(Z')$.

Let Z be the union of all the q-cells in Z' with compact closure in Z'. This is a CW-complex. Hamm shows that $\Psi^{-1}(Z)$ is a strong deformation retract of $\Psi^{-1}(Z')$ [Hamm$_2$, Lemma 3.2].

For all j with $1 \leq j \leq n+2$ the hyperplanes $w_i = 0$ for $i \neq j$ bound a simplex σ_j in Z (see Figure 2.3.1). Let $\varepsilon_j := \tilde{\Psi}(\sigma_j)$ be the corresponding simplex in $\tilde{\Psi}(Z)$. For $\kappa \in \{0,1\}$ we define

$$e_{2j-1+\kappa} := \tau_j^{\kappa}(1-\tau_1)\ \dots\ (\widehat{1-\tau_j})\ \dots\ (1-\tau_{n+2})\varepsilon_j .$$

One can easily show that these elements are cycles in $\Psi^{-1}(Z)$.

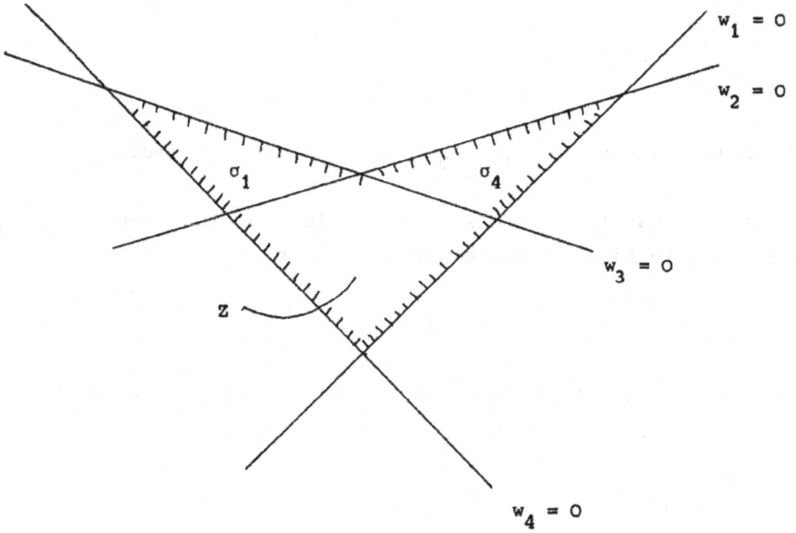

Figure 2.3.1: Picture for $n=2$.

Proposition 2.3.1. (i) The elements $e_1, e_2, \ldots, e_{2n+3}$ represent a basis of $H_n(\Psi^{-1}(Z), \mathbb{C}) = H_n(X_s, \mathbb{C}) = H_{\mathbb{C}}$.

(ii) One has

$$e_{2n+4} = e_{2n+3} + \sum_{j=1}^{n+1} (-1)^{n-j} (e_{2j-1} - e_{2j}) .$$

(iii) Let $h_* : H_{\mathbb{C}} \longrightarrow H_{\mathbb{C}}$ be the monodromy operator induced by the loop $\tilde{\omega} : S^1 \longrightarrow \mathbb{C}^2$, $\varphi \longrightarrow (0, \eta \exp(2\pi\sqrt{-1}\varphi))$. Then h_* coincides with the mapping $\tau_1 \tau_2 \cdots \tau_{n+2}$.

Proof. Assertion (i) is Lemma 3.6 and assertion (iii) is Lemma 4.1 of [Hamm$_2$]. We prove (ii). We show by induction on n

$$\varepsilon_{n+2} = \sum_{j=1}^{n+1} (-1)^{j+1} \varepsilon_{n+2-j} . \qquad (2.3.2)$$

From this formula one can easily derive the formula of (ii). For the case $n=1$ one has the following picture according to the order between the numbers a_j :

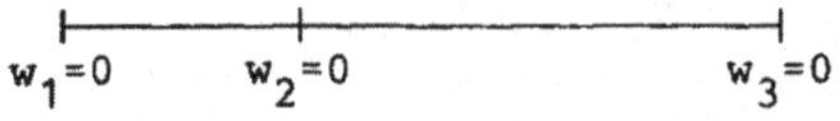

Figure 2.3.2

From this figure one gets $\varepsilon_3 = \varepsilon_2 - \varepsilon_1$, which is formula (2.3.2) for $n = 1$.

Now assume that the formula is valid for n. The simplex ε_{n+3} is a linear combination of the other simplices ε_j ,

$$\varepsilon_{n+3} = \sum_{j=1}^{n+2} k_j \varepsilon_j \quad , \quad k_j \in \mathbb{Z} \quad .$$

The hyperplane $w_{n+3} = 0$ bounds all the simplices ε_j with $1 \leq j \leq n+2$, but ε_{n+3}. Therefore

$$\sum_{j=1}^{n+2} k_j \varepsilon_j \cap \{w_{n+3} = 0\} = \phi \ .$$

By the induction assumption for $Z' \cap \{w_{n+3} = 0\}$, one has

$$\varepsilon_{n+3} = k_{n+2}\left(\varepsilon_{n+2} - \sum_{j=1}^{n+1} (-1)^{j+1} \varepsilon_{n+2-j}\right) \quad .$$

The assumption $a_1 < a_2 < \ldots < a_{n+2}$ implies $\varepsilon_{n+3} \subset \varepsilon_{n+2}$. This proves $k_{n+2} = 1$, and hence formula (2.3.2) for $n+1$. This completes the proof of Proposition 2.3.1.

Corollary 2.3.3. The action of the monodromy operator h_* on the cycles e_i can be described as follows:

$$\left.\begin{array}{ll} h_*(e_{2j-1}) & = (-1)^{n+1} e_{2j} \\ h_*(e_{2j}) & = (-1)^{n+1} e_{2j-1} \end{array}\right\} \quad \text{for} \quad 1 \leq j \leq n+2 \ .$$

This corollary is derived from the preceding proposition by a simple calculation.

We shall now construct vanishing cycles δ_i which can be transformed into the cycles e_i. The critical locus of f consists of the coordinate axes in $\mathbb{C}^{n+2}$. The discriminant locus D_f of f consists of $n+2$ different double lines through the origin in $\mathbb{C}^2$. Let $t > 0$ be a small real number and $\bar{D} = \{y \in \mathbb{C} \mid |y| \leq \eta\}$. We consider the disc $\{t\}\times\bar{\mathbb{D}}$ and choose the point $s=(t,\eta)$ as the base point. For sufficiently small t, the disc intersects the discriminant D_f in $n+2$ different points $\tilde{s}_j=(t,a_jt)$, $j=1,\ldots,n+2$, lying on the real axis in the interior of $\bar{\mathbb{D}}$. The points are numbered according to their order on the real axis. Let ρ be a positive real number, which is small enough such that a_jt is the only point of D_f in the interval $[a_jt-\rho, a_jt+\rho]$. We choose a strongly distinguished system of paths $(\tilde{\varphi}_1,\ldots,\tilde{\varphi}_{n+2})$ joining the points $\tilde{s}_j$ with s, by combining each interval $[a_jt, a_jt+\rho]$ with a path from $a_jt+\rho$ to η as in Figure 2.3.3.

The fibre over the point $\tilde{s}_j$, $j=1,\ldots,n+2$, has exactly two ordinary double points x_{2j-1} and x_{2j} given by $z_j=\pm\sqrt{t}$, $z_i=0$ for $i\neq j$. Then the above system of paths $(\tilde{\varphi}_1,\ldots,\tilde{\varphi}_{n+2})$ determines (up to orientation) a strongly distinguished system of generators $(\delta_1,\ldots,\delta_{2n+4})$ of vanishing cycles in $H_n(X_s)$. More precisely, one has to take a nearby Morse function with distinct critical values $s_1,s_2,\ldots,s_{2n+4}$, and paths φ_{2j-1}, φ_{2j} adjacent to the paths $\tilde{\varphi}_j$ as indicated in Figure 2.3.3 by dotted lines. This strongly distinguished system of generators $(\delta_1,\ldots,\delta_{2n+4})$ has the following properties:

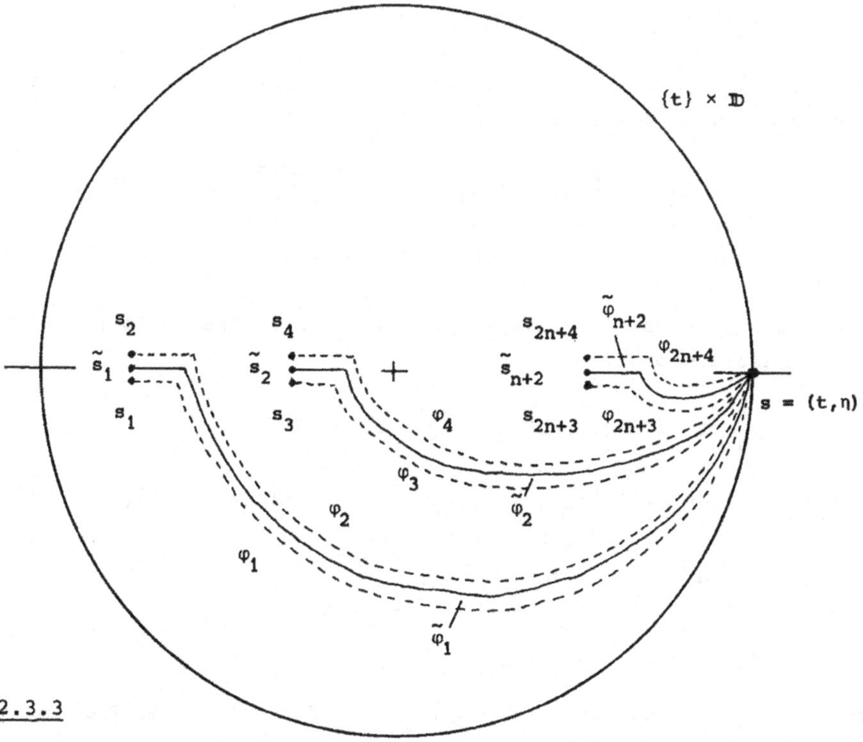

Figure 2.3.3

Proposition 2.3.4. If t tends to 0, then the vanishing cycle $\delta_i \in H_n(X_s)$, $i = 1,\dots,2n+4$, passes to the cycle $\pm e_i \in H_n(X_{s_0})$. In other words, $(e_1,\dots,e_{2n+4})$ is a strongly distinguished system of generators of vanishing cycles of the singularity $(X,0)$. The corresponding Coxeter element $c = \gamma_1\gamma_2\cdots\gamma_{2n+4}$ coincides with the monodromy operator h_*.

Proof. We derive the equations for the vanishing cycles δ_i according to Chapter 1.2. Consider the critical point x_{2j-1} given by $z_j = +\sqrt{t}$, $z_i = 0$ for $i \neq j$. In a neighbourhood B_{2j-1} of x_{2j-1} we make the coordinate transformation

$$u_j = 2\sqrt{t}(z_j - \sqrt{t}) + \sum_{i=1, i\neq j}^{n+2} z_i^2 + (z_j - \sqrt{t})^2 ,$$

$$u_i = \sqrt{a_i - a_j}\, z_i \quad \text{for } i \neq j.$$

Then the fibre $X_t' = f_1^{-1}(t) \cap B_\varepsilon$ is locally given by the equation $u_j = 0$, and one can take u_i, $i \neq j$, $1 \leq i \leq n+2$, as local coordinates around the point x_{2j-1}. Then $f|_{B_{2j-1}}$ is given by

$$f(u) = a_i t + u_1^2 + \dots + u_{j-1}^2 + u_{j+1}^2 + \dots + u_{n+2}^2$$

The vanishing cycle in the fibre $Y_* = f^{-1}(\tilde{s}_j + \rho) \cap B_{2j-1}$ over $\tilde{s}_j + \rho$ is represented by the sphere

$$S^n = \{u \in \mathbb{C}^{n+2} \mid u_j = 0;\ u_1^2 + \dots + u_{j-1}^2 + u_{j+1}^2 + \dots + u_{n+2}^2 = \rho;$$

$$\operatorname{Im} u_i = 0 \quad \text{for } i \neq j\} .$$

In this coordinates z_i this sphere is given by the following equations (for ρ small enough):

$$z_j^2 = t - z_1^2 - \dots - z_{j-1}^2 - z_{j+1}^2 - \dots - z_{n+2}^2, \tag{2.3.5}$$

$$(a_1-a_j)z_1^2 + \dots + (a_{j-1}-a_j)z_{j-1}^2 + (a_{j+1}-a_j)z_{j+1}^2 + \dots + (a_{n+2}-a_j)z_{n+2}^2 = \rho, \tag{2.3.6}$$

$$\sqrt{a_i - a_j}\, z_i \in \mathbb{R} \quad \text{for } i \neq j , \tag{2.3.7}$$

$$z_j > 0. \tag{2.3.8}$$

Here (2.3.7) implies that $(a_i - a_j)z_i^2 \in \mathbb{R}_+$ for all $i \neq j$. Then (2.3.5) implies that $z_j^2 \in \mathbb{R}$.

Now let $\tilde{\varphi}'_j(r) = \tilde{\varphi}_j(r) - a_j t$ for $r \in [0,1]$. Then transport along $\tilde{\varphi}_j$ yields a sphere in the fibre over $\tilde{\varphi}_j(r)$ given by analogous equations as above, where one has to replace ρ by $\tilde{\varphi}'_j(r)$ in (2.3.6), and $\mathbb{R}$ by $\mathbb{R}\ \exp(\sqrt{-1}\ \arg(\tilde{\varphi}'_j(r)))$, and to choose the root of z_j^2 in (2.3.5) by analytic continuation of (2.3.8). In the fibre $X_s = f^{-1}(s) \cap B_\varepsilon$ over $s = (t,\eta)$, this sphere is given by the equations (2.3.5), (2.3.6) with ρ replaced by $\eta - a_j t$, (2.3.7) and

$$z_j > 0 \quad \text{or} \quad -\sqrt{-1}z_j > 0 \tag{2.3.9}$$

After the passage $t \longrightarrow 0$ one obtains the set E_{2j-1} given by the corresponding equations with $t = 0$.

<u>Assertion 2.3.10.</u> $\Psi(E_{2j-1}) = \sigma_j$.

It follows from (2.3.7) that $(a_i - a_j)z_i^2 \in \mathbb{R}_+$ for all $i \neq j$. But this means that $z_i^2 \in \mathbb{R}$ and $\operatorname{sign} z_i^2 = \operatorname{sign}\ (a_i - a_j) = \operatorname{sign}\ (i - j)$ or $z_i^2 = 0$. Therefore

$$\Psi(E_{2j-1}) = \{w \in Z' \mid \operatorname{sign} w_i = \operatorname{sign}(i-j) \quad \text{for} \quad i \neq j \quad \text{or} \quad w_i = 0\} .$$

We show $\Psi(E_{2j-1}) = \sigma_j$ by induction on n. For $n = 1$ a simple calculation yields the following picture

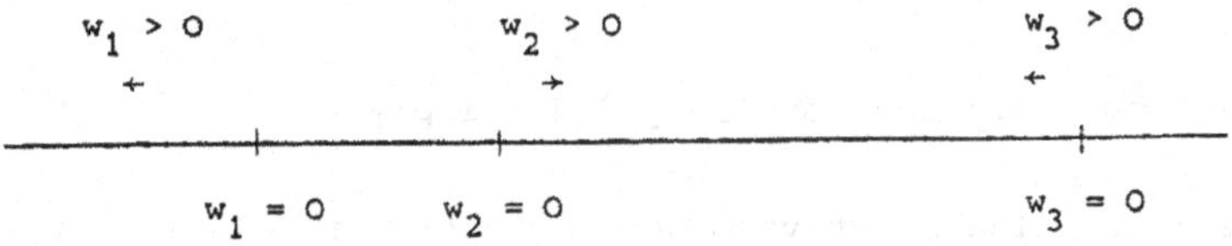

<u>Figure 2.3.4</u>

This proves the assertion for $n = 1$. Now assume that the assertion is true for $n - 1$. Then one has for all $i \neq j$

$$\Psi(E_{2j-1}) \cap \{w_i = 0\} = \sigma_j \cap \{w_i = 0\} .$$

But the union of the simplices on the righthand side is the boundary of σ_j, the complement of which in Z' has two connected components. One of these components, namely the interior of σ_j, is convex. Since $\Psi(E_{2j-1})$ is convex, assertion (2.3.10) follows.

Property (2.3.9) and the definition of e_{2j-1} therefore imply

$[E_{2j-1}] = \pm e_{2j-1}$.

An analogous consideration for the critical point x_{2j} yields the cycle $\pm e_{2j}$. This proves the first part of the proposition.

The second part of the proposition follows from the fact that the loops defining the corresponding operators are homotopic.

Remark 2.3.11. The thimble $\hat{\delta}_{2j-1+\kappa}$ ($\kappa \in \{0,1\}$) corresponding to the vanishing cycle $\delta_{2j-1+\kappa}$ passes, for $t \longrightarrow 0$, to the set

$$\tau_j^{\kappa}(1-\tau_1)\dots\widehat{(1-\tau_j)}\ \dots\ (1-\tau_{n+2})(\mathrm{cone}(\sigma_j)),$$

where $\mathrm{cone}(\sigma_j)$ is the real cone over σ_j with apex 0.

We orientate the cycles δ_i in such a way that

$$\delta_{2j-1} = (-1)^{n+1} e_{2j-1},$$

$$\delta_{2j} = e_{2j}$$

for $j = 1,\dots,n+2$. We set

$$r = \sum_{j=1}^{n+2} (-1)^{n-j}(e_{2j-1} - e_{2j})$$

$$= \sum_{j=1}^{n+2} (-1)^{n-j}((-1)^{n+1}\delta_{2j-1} - \delta_{2j}).$$

Then Proposition 2.3.4 and Corollary 2.3.2 imply

Corollary 2.3.12. The relative monodromy operator $\hat{c}$ of $(X,0)$ is described by the following equations:

$$\hat{c}(\hat{\delta}_{2j-1}) = \hat{\delta}_{2j} + b_{2j-1}r,$$

$$\hat{c}(\hat{\delta}_{2j}) = \hat{\delta}_{2j-1} + b_{2j}r$$

for $j = 1,\dots,n+2$ and certain numbers $b_{2j-1}, b_{2j} \in \mathbb{Z}$.

Using these results we can determine the intersection matrix of the cycles e_i, which was not done by Hamm. The self-intersection numbers of the cycles e_i are determined by Proposition 2.3.4 (cf. Chapter 1.2). Using the Picard-Lefschetz formulas or Proposition 1.6.3 one can compute the other intersection numbers of the $\hat{\delta}_i$, and hence of the e_i, from Corollary 2.3.12. Such a calculation will be demonstrated for a more complicated example in the next section. Therefore we omit the details of the calculation in this case and only indicate the result.

Corollary 2.3.13. The intersection matrix of the strongly distinguished basis $(\hat{\delta}_1,\ldots,\hat{\delta}_{2n+4})$ of thimbles is given by the following formulas

$$\langle\hat{\delta}_i,\hat{\delta}_j\rangle = \begin{cases} 0 & \text{for } j = i+1,\ j \text{ even}, \\ \left.\begin{cases} (-1)^{n/2} & \text{for } n \text{ even}, \\ (-1)^{(n+1)/2}(-1)^{i+j} & \text{for } n \text{ odd}, \end{cases}\right\} & \text{for } i<j \text{ otherwise}. \end{cases}$$

Hence, disregarding the weights of the edges, the Dynkin diagram corresponding to the basis $(\hat{\delta}_1,\ldots,\hat{\delta}_{2n+4})$ is a complete graph with $2n+4$ vertices, where the edges between the vertices corresponding to $\hat{\delta}_{2j-1}$ and $\hat{\delta}_{2j}$ for $j = 1,\ldots,n+2$ are omitted.

Remark 2.3.14. The thimbles $\hat{\delta}_i$ were constructed by the choice of a certain strongly distinguished system of paths $(\tilde{\varphi}_1,\ldots,\tilde{\varphi}_{n+2})$ from the points $\tilde{s}_1,\ldots,\tilde{s}_{n+2}$ to the base point s. Corollary 2.3.13 implies that each strongly distinguished system of $n+2$ paths from the points $\tilde{s}_1,\ldots,\tilde{s}_{n+2}$ to s defines in the same way a strongly distinguished basis of thimbles, its intersection matrix being equal to that of $(\hat{\delta}_1,\ldots,\hat{\delta}_{2n+4})$ after an appropriate orientation of the thimbles. This follows from the fact that the action of a generator $\tilde{\alpha}_j$, $j = 1,\ldots,n+1$, of the braid group Z_{n+2} on the systems of paths $(\tilde{\varphi}_1,\ldots,\tilde{\varphi}_{n+2})$,

$$\tilde{\alpha}_j : (\tilde{\varphi}_1,\ldots,\tilde{\varphi}_{n+2}) \rightarrow (\tilde{\varphi}_1,\ldots,\tilde{\varphi}_{j-1},\tilde{\varphi}_{j+1}\omega_j,\tilde{\varphi}_j,\tilde{\varphi}_{j+2},\ldots,\tilde{\varphi}_{n+2}),$$

induces the following transformation of the basis $(\hat{\delta}_1,\ldots,\hat{\delta}_{2n+4})$:

$$\alpha_{2j}\,\alpha_{2j+1}\,\alpha_{2j-1}\,\alpha_{2j} \quad .$$

Because of the intersection matrix of $(\hat{\delta}_1,\ldots,\hat{\delta}_{2n+4})$, this transformation coincides with

$$\kappa_{2j-1}\,\kappa_{2j} \qquad ,$$

hence with two changes of orientations. Since the braid group Z_{n+2} acts transitively on the set of all strongly distinguished systems of $n+2$ paths from the points $\tilde{s}_1,\ldots,\tilde{s}_{n+2}$ to s, the assertion follows.

If we admit in the definition of f pairwise distinct complex numbers a_j, then the discriminant D_f intersects the disc $\{t\}\times\bar{\mathbb{D}}$ also in $n+2$ different points $\tilde{s}_j = (t,a_jt)$, which now no longer lie on the real axis. Each strongly distinguished system of paths of the points $\tilde{s}_1,\ldots,\tilde{s}_{n+2}$ to s now yields as above a strongly distinguished basis of thimbles having the intersection matrix of Corollary 2.3.13 with an appropriate orientation. This follows from the fact that one can choose $n+2$ analytic functions $g_j:[0,1]\longrightarrow\bar{\mathbb{D}}$, $j=1,\ldots,n+2$, with $g_j(0)=a_j$, $g_j(1)\in\mathbb{R}\cap\bar{\mathbb{D}}$, and $g_i(\theta)\neq g_j(\theta)$ for $1\leq i\neq j\leq n+2$, $\theta\in[0,1]$, and consider the deformation of f obtained in this way.

We now analyse the Milnor lattice $H=\hat{H}/\mathbb{Z}\cdot r$, first in the skew-symmetric case (dimension n odd). By the transformations

$$\beta_{2n+3}(2j),\ \beta_{2j-1}(2j) \qquad \text{for } j=1,\ldots,n+1\ ;$$

$$\beta_{2j+1}(2j-1) \qquad \text{for } j=1,\ldots,n+1\ ;$$

the basis $(\hat{\delta}_1,\ldots,\hat{\delta}_{2n+4})$ is transformed to a weakly distinguished basis of thimbles with the following intersection matrix. The intersection matrix reduced modulo 2 is described by the Dynkin diagram of Figure 2.3.5.

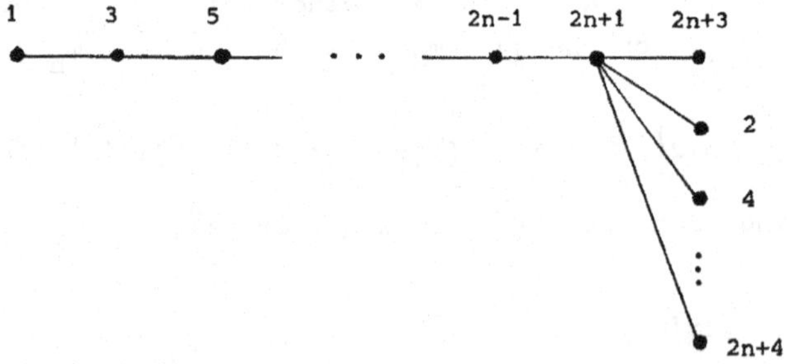

Figure 2.3.5

By [Janssen$_1$, (3.10)] this basis is equivalent to a special weakly distinguished basis in the sense of [Wajnryb$_1$]. Therefore the skew-symmetric vanishing lattice $(\hat{H},\hat{\Delta})$ (respectively (H,Δ)) of an odd-dimensional complete intersection of two quadrics with an isolated singularity is the vanishing lattice

$$A^{odd}(\underbrace{1,\ldots,1}_{n+1/2};p;0)$$

with $p = n+3$ (respectively $p = n+2$) in Janssen's classification [Janssen$_2$].

For the investigation of the even-dimensional case we transform the strongly distinguished basis $(\hat{\delta}_1,\ldots,\hat{\delta}_{2n+4})$ of thimbles using the braid group action of Chapter 1.5. The transformations

$$\text{(A)}\quad \alpha_2,\alpha_3,\ldots,\alpha_{2n+3} \quad\text{and}\quad \begin{cases}\kappa_{2n+4} & \text{if } n \text{ is even,}\\ \kappa_1 & \text{if } n \text{ is odd ,}\end{cases}$$

transform the basis $(\hat{\delta}_i)$ to a strongly distinguished basis $(\hat{\delta}_i')$ of thimbles with

$$\langle\hat{\delta}_i',\hat{\delta}_j'\rangle = \begin{cases} 0 & \text{for } j=i+1,\ j \text{ odd},\ (i,j)=(1,2n+4)\\ \left.\begin{cases}(-1)^{n/2} & \text{for } n \text{ even}\\ (-1)^{(n+1)/2}(-1)^{i+j} & \text{for } n \text{ odd}\end{cases}\right\} & \text{for } i<j \text{ otherwise.}\end{cases}$$

(2.3.15)

We apply the following transformations to the basis $(\hat{\delta}_1',\ldots,\hat{\delta}_{2n+4}')$:

(B) $\alpha_{2n-1}, \alpha_{2n}$;

$\alpha_{2n-3}, \alpha_{2n-2}, \alpha_{2n-1}, \alpha_{2n}$;

.

.

.

$\alpha_1, \alpha_2,\ldots, \alpha_{2n-1}, \alpha_{2n}$;

$\alpha_{2n+3}, \alpha_{2n+2}$.

For $n=1$, we denote the basis obtained in this way by $(\hat{\delta}_1'',\ldots,\hat{\delta}_6'')$. The corresponding Dynkin diagram is shown in Figure 2.3.6.

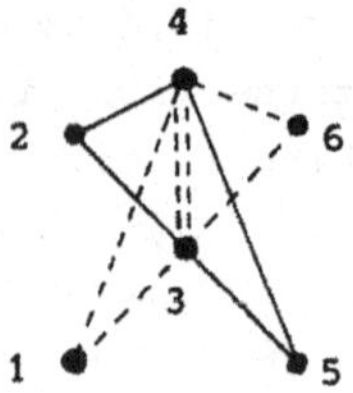

Figure 2.3.6

If $n \geq 2$ then we perform further transformations, namely the transformations

(C) $\beta_n, \beta_{n+1}; \kappa_{2n+2}, \kappa_{2n+3}, \kappa_{2n+4}$.

For $n=2$, we denote the basis obtained in this way by $(\hat{\delta}_1'',\ldots,\hat{\delta}_8'')$. Finally, in the case $n \geq 3$, we perform the following additional transformations:

(D) $\beta_{n-1}, \beta_{n-2},\ldots, \beta_2;$
$\beta_n, \beta_{n-1},\ldots, \beta_3;$
$\beta_{n+1}, \beta_{n+2}; \beta_n, \beta_{n+1};\ldots;\beta_4, \beta_5$.

Thus we obtain a basis of thimbles, which is still strongly distinguished, and which will be denoted by $(\hat{\delta}_1'',\ldots,\hat{\delta}_{2n+4}'')$. The Dynkin diagram corresponding to this basis is shown in Figure 2.3.7 (for n even; for n odd one still has to apply the changes of orientations κ_1, κ_6 for $n=1$, respectively $\kappa_2, \kappa_4; \kappa_{n+3}, \kappa_{n+4},\ldots,\kappa_{2n}; \kappa_{2n+1}, \kappa_{2n+4}$ for $n \geq 3$, in advance).

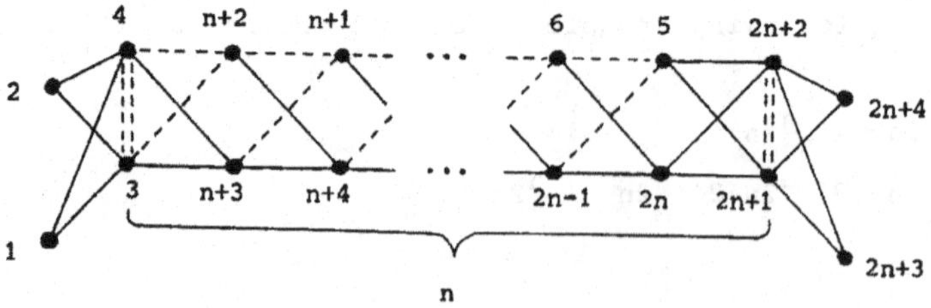

Figure 2.3.7

2.4 Extended affine root systems

The graph of Figure 2.3.7 is very similar to the Dynkin diagram of an extended affine root system of type $D_{n+3}^{(1,1)}$ given by K. Saito in [Saito$_1$, p. 122]. This graph is shown in Figure 2.4.1

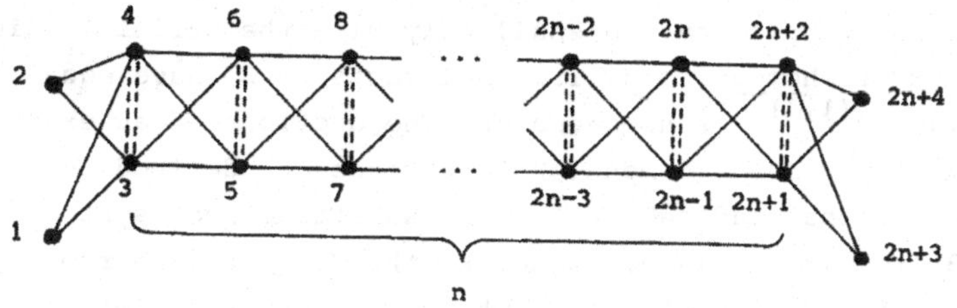

Figure 2.4.1

We briefly explain the meaning of this graph (cf. [Saito$_1$]). Let V be an $\mathbb{R}$-vector space with a symmetric bilinear form $I: V \times V \longrightarrow \mathbb{R}$ of signature (t_+, t_0, t_-). Here t_+ (respectively t_-) is the dimension of a maximal positive (respectively negative) definite subspace of V with respect to I, and t_0 is the dimension of the kernel of I. For $\lambda \in V$ with $I(\lambda,\lambda) \neq 0$ we define $\lambda^\vee \in V$ and $w_\lambda \in GL(V)$ as follows:

$$\lambda^\vee := 2\lambda / I(\lambda,\lambda) ,$$

$$w_\lambda(v) := v - I(v,\lambda^\vee)\lambda \quad \text{for} \quad v \in V.$$

Definition 2.4.1. A subset $R \subset V$ is called an extended affine root system, if the following conditions are satisfied:

(i) $(t_+, t_0, t_-) = (\ell, 2, 0)$

(ii) $V = Q(R) \otimes_{\mathbb{Z}} \mathbb{R}$ for $Q(R) = \mathbb{Z} \cdot R$.

(iii) $I(\lambda,\lambda) \neq 0$ for all $\lambda \in R$.

(iv) $w_\lambda R = R$ for all $\lambda \in R$.

(v) $I(\lambda, \tilde{\lambda}^\vee) \in \mathbb{Z}$ for all $\lambda, \tilde{\lambda} \in R$.

(vi) (Irreducibility) If $R = R_1 \dot{\cup} R_2$ with $R_1 \subset R_2^\perp$, $R_2 \subset R_1^\perp$ (with respect to I), then $R_1 = \phi$ or $R_2 = \phi$.

Let W_R be the subgroup of $O(V)$ generated by the w_λ for $\lambda \in R$.

K. Saito has classified all extended affine root systems. This classification is closely related to the corresponding classification in the classical case, namely the case where I is positive definite. In particular there is (up to isomorphism) only one extended affine root system such that its quotient by the radical is the classical root system D_ℓ , and this is denoted by Saito by $D_\ell^{(1,1)}$. The upper index has no meaning for our consideration, and we shall not explain it.

Saito defines a Dynkin diagram for an extended affine root system as follows. For simplicity, we shall only give the definition in the case where R is <u>homogeneous</u>, i.e. all roots of R have equal length. In particular $D_\ell^{(1,1)}$ is homogeneous. The quotient of an extended affine root system R by any 1-dimensional subspace K defined over $\mathbb{Q}$ of the 2-dimensional radical $\ker I$ is an affine root system [Saito$_1$, (3.1)]. If R is homogeneous, then this affine root system is uniquely determined up to isomorphism [Saito$_1$, §5]. Let $\{\lambda_0,\lambda_1,\dots,\lambda_\ell\}$ be a basis of this affine root system such that $\{\lambda_1,\dots,\lambda_\ell\}$ is a (Weyl-) basis of the corresponding classical root system, and

$$\lambda_0 = \sum_{i=1}^{\ell} m_i\lambda_i, \quad m_i > 0 ,$$

is the longest positive root. We set $m_0 = 1$. Let a be a generator of K over $\mathbb{Q}$. We define for an index j with $0 \leq j \leq \ell$ and $m_j = \max_i \{m_i\}$

$$\lambda_j^* = \lambda_j + a .$$

<u>Definition 2.4.2.</u> The <u>Dynkin diagram</u> G_R of an extended affine root system R is the Dynkin diagram corresponding to $\{\lambda_0,\lambda_1,\dots,\lambda_\ell\} \cup \{\lambda_j^* \mid m_j = \max_i \{m_i\}\}$.

For $D_\ell^{(1,1)}$ one just obtains the graph of Figure 2.4.1 as a Dynkin diagram. We denote the set of vertices of G_R by $|G_R|$. Then one has the following facts:

(a) $\# \ |G_R| > \dim V$

(b) The lattice $Q(R)$, the group W_R, and R are determined by G_R in the usual way:

$$Q(R) = \sum_{\lambda\in|G_R|} \mathbb{Z}.\lambda \ , \ W_R = \langle w_\lambda \mid \lambda \in G_R\rangle, \ R = \bigcup_{\lambda\in|G_R|} W_R.\lambda \ .$$

(c) One can reconstruct V and R from G_R as follows: Let $\hat{V}$ be the $\mathbb{R}$-vector space of dimension $\#|G_R|$ spanned by the

$\lambda \in |G_R|$, and let $\hat{Q}$ be the corresponding $\mathbb{Z}$-lattice. To avoid confusions, we denote the basis element of $\hat{V}$ corresponding to $\lambda \in |G_R|$ by $\hat{\lambda}$. Let $\hat{I}$ be the bilinear form defined by the Dynkin diagram G_R, and let

$$\hat{W} := \langle w_{\hat{\lambda}} \mid \lambda \in |G_R| \rangle$$

$$\hat{R} := \bigcup_{\lambda \in |G_R|} \hat{W}.\hat{\lambda} .$$

The product $\hat{c}_R \in \hat{W}$ of the $w_{\hat{\lambda}_p}$ with $m_p \neq \max_i \{m_i\}$ and of the products $w_{\hat{\lambda}_j} w_{\hat{\lambda}_j^*}$ for $0 \leq p, j \leq \ell$ in any order is called a <u>pre-Coxeter element</u> of R. By [Bourbaki$_2$, Chap. V, §6, Lemma 1] any two pre-Coxeter elements are conjugate in $\hat{W}$. By [Saito$_1$, (9.6)] one has the following proposition.

<u>Proposition 2.4.3. Let</u> $\hat{c}_R = \hat{c}_{R,s}\,\hat{c}_{R,u}$ <u>be the (multiplicative) Jordan decomposition of</u> $\hat{c}_R$ <u>into a semi-simple part</u> $\hat{c}_{R,s}$ <u>and a unipotent part</u> $\hat{c}_{R,u}$. <u>Then one has</u>

$$V = \hat{V}/\mathrm{im}\,(\hat{c}_{R,u} - \mathrm{id}),$$
$$Q = \hat{Q}/\mathrm{im}\,(\hat{c}_{R,u} - \mathrm{id}) \cap \hat{Q},$$

<u>and</u> R <u>is the image of</u> $\hat{R}$ <u>in</u> V.

We now compare the graphs of Figure 2.3.7 and Figure 2.4.1. The two graphs coincide in the case $n = 2$. For $n \geq 3$, the following transformations, which have to be applied to the graph of Figure 2.3.7, show that the two graphs are weakly equivalent:

$$\alpha_{n+1}(n+2), \alpha_{n+4}(n+2); \alpha_n(n+1), \alpha_{n+5}(n+1); \ldots; \alpha_5(6), \alpha_{2n}(6);$$

$$\alpha_{2n+3}(2n+1), \alpha_{2n+4}(2n+1), \alpha_{2n+1}(5), \alpha_{2n+2}(5), \alpha_{2n+3}(2n+1), \alpha_{2n+4}(2n+1);$$

$$\kappa_5, \kappa_6, \ldots, \kappa_{n+2} .$$

This implies that we have the following isomorphisms, setting $R = D_{n+3}^{(1,1)}$:

$$\hat{H} \cong (\hat{Q}(R), \varepsilon\hat{I}) \cong D_{n+3} \perp \ker \varepsilon\hat{I},$$
$$\dim(\ker \varepsilon\hat{I}) = n+1,$$
$$\hat{\Delta} \cong \hat{R},$$

$$\hat{\Gamma} \cong \hat{W}_R ,$$

where $\hat{H}, \hat{\Delta}$, and $\hat{\Gamma}$ are the corresponding invariants of an isolated singularity $(X,0)$ of an n-dimensional complete intersection of two quadrics, n is even and $\varepsilon = (-1)^{n/2}$.

This implies in particular that the set of vanishing cycles Δ of $(X,0)$ contains a root system of type D_{n+3}, which was already known (cf. [Deligne-Katz, Exposé XIX, Proposition 5.2 (iii)] [Reid_1]), and has led to the notation $\tilde{D}_{n+3}$ for these singularities (cf.[Knörrer_1]).

In addition in the case $n = 2$, the following is true for the relative monodromy operator $\hat{c}$ and the vanishing lattice (H,Δ) of $(X,0)$:

$$\hat{c} = \hat{c}_R, \quad H = (Q(R),-I), \quad \Delta = R.$$

It is, however, not true, as I unfortunately claimed erroneously in a letter to K. Saito (cf. [Saito_1, Added in Proof]), that the two graphs are strongly equivalent for $n \geq 3$. This follows from the fact that the corresponding Coxeter elements $\hat{c}$ and $\hat{c}_R$ are different: By Proposition 1.6.6 (n even) one has

$$\dim(\operatorname{im}(\hat{c}_u - \mathrm{id})) = 1 \quad ,$$

whereas

$$\dim(\operatorname{im}(\hat{c}_{R,u} - \mathrm{id})) = \ell - 4 = n - 1$$

by Proposition 2.4.3. This difference is due to the fact that the lattices H and $Q(R)$ are different except for $n = 2$:

$$\dim(\ker H) = n, \text{ but } \dim(\ker Q(R)) = 2.$$

But note the analogy of Proposition 1.6.6 and Proposition 2.4.3. Also compare Proposition 3.5.2.

2.5. A generalization of a method of Lazzeri

Unfortunately, the application of the method described in Section 2.2 already leads to difficult calculations for somewhat more complicated singularities. This applies in particular to an example which will play an important rôle for us. Therefore we want to present another method for computing Dynkin diagrams by means of this example. We shall generalize a method of Lazzeri in the hypersurface case (see [Hefez-Lazzeri]).

We shall briefly describe this method. Let $f:(\mathbb{C}^{n+2},0) \longrightarrow (\mathbb{C}^2,0)$ be the germ of an analytic mapping with an isolated singularity at 0, and let $f : X_P \longrightarrow P$ be a suitable representative. We assume that the discriminant D_f of f is reduced and that the multiplicity of D_f at 0 coincides with the multiplicity m of the discriminant D_F of the semi-universal deformation. We now carry out the same construction as in Chapter 1.1 with the only difference that we consider the target $P \subset \mathbb{C}^2$ of the mapping f and not the base space of the semi-universal deformation. Hence we choose P of the form $P = T \times \mathbb{D}$ for two discs T and $\mathbb{D}$ such that $\mathbb{D} = \ell \cap P$ for a line ℓ not contained in the tangent cone of D_f at 0. Let $t \in T$ be a point of T which is not contained in the image of the ramification locus of the projection $\pi|D_f : D_f \longrightarrow T$. We set $\mathbb{D}^* = (\{t\} \times \mathbb{D}) - ((\{t\} \times \mathbb{D}) \cap D_f)$ and choose a base point $s \in \mathbb{D}^*$.

Let $(\varphi_1,\ldots,\varphi_m)$ be a strongly distinguished system of paths in $\{t\} \times \mathbb{D}$ from the intersection points with D_f to s. We determine the intersection matrix of a corresponding strongly distinguished basis $(\hat{\delta}_1,\ldots,\hat{\delta}_m)$ of thimbles as follows: The system $(\omega_1,\ldots,\omega_m)$ of simple loops corresponding to the system of paths $(\varphi_1,\ldots,\varphi_m)$ generates the fundamental group $\pi_1(P-(P \cap D_f),s)$. The corresponding relations can be deduced from a theorem of van Kampen. By means of the Picard-Lefschetz formulas one can derive from these relations relations between the corresponding Picard-Lefschetz transformations γ_i , $i = 1,\ldots,m$. From these relations one can obtain equations for the intersection numbers. Compared with the hypersurface case, there is, however, an additional difficulty due to the fact that there are linear relations between the vanishing cycles δ_i which yield additional unknowns for the equations. We now consider an example which can nevertheless be treated successfully by this procedure.

The example considered by us is the n-dimensional singularity given by the mapping $f : \mathbb{C}^{n+2} \longrightarrow \mathbb{C}^2$, n even, with

$$f = g \oplus g' \oplus h^{(1)} \oplus \ldots \oplus h^{((n-2)/2)} ,$$

$$g(z_1,z_2) = (z_1^2 + z_2^3 + z_1z_2, z_1z_2) ,$$

$$g'(z_3,z_4) = (z_3z_4, z_3^2 + z_4^3),$$

$$h^{(q)}(z_{2q+3},z_{2q+4}) = (z_{2q+3}z_{2q+4}, a_{2q-1}z_{2q+3}^2 + a_{2q}z_{2q+4}^2 - z_{2q+3}^2 z_{2q+4}^2)$$

$$q = 1,\ldots,(n-2)/2;\ a_j \in \mathbb{R},\ 1 < a_1 < a_2 < \ldots < a_{n-2}\ .$$

For the definition of $\oplus$ see Chapter 3.1. We shall introduce for this singularity in Chapter 3 the notation T^n_{2233}.

By [Wall$_2$, Lemma 1.4] the discriminant D_f of f can be decomposed as follows:

$$D_f = D_g \cup D_{g'} \cup D_{h^{(1)}} \cup \ldots \cup D_{h^{((n-2)/2)}}\ .$$

Let (u,v) be coordinates of P with $f_1(z) = u$, $f_2(z) = v$. Then one easily calculates that D_g and $D_{g'}$ can be described in these coordinates as follows

$$D_1 = D_g\ :\ (u-v)^5 - (5^5/3^3 2^2)v^6 = 0,$$

$$D_2 = D_{g'}:\ u^6 \quad - (3^3 2^2/5^5)v^5 = 0.$$

The curve $D_{h^{(q)}}$ has two branches. Their Puiseux parametrizations have the following beginning terms:

$$D_{2+q} = D^{(1)}_{h^{(q)}}\ :\ v = 2a_{2q}\sqrt{\frac{a_{2q-1}}{a_{2q}}}\ u - \sqrt[4]{\frac{a_{2q-1}}{a_{2q}}}\ u^{3/2} + \ldots\ ,$$

$$D_{(n/2)+1+q} = D^{(2)}_{h^{(q)}}\ :\ v = -2a_{2q}\sqrt{\frac{a_{2q-1}}{a_{2q}}}\ u - \sqrt{-1}\sqrt[4]{\frac{a_{2q-1}}{a_{2q}}}\ u^{3/2} + \ldots\ ,$$

We choose

$$T = \{(u,v) \in \mathbb{C}^2 \mid v = 0,\ |u| \leq \eta_1\},$$

$$\mathbb{D} = \{(u,v) \in \mathbb{C}^2 \mid u = 0,\ |v| \leq \eta_2\},$$

where $\eta_1 \ll \eta_2$ are chosen small enough. We choose the base point $t = (\eta_1, 0) \in T$ and in $\{t\} \times \mathbb{D}$ the base point $s = (\eta_1, \eta_2)$. We now consider a strongly distinguished system of paths $(\varphi_1,\ldots,\varphi_{2n+6})$ from the $m = 2n+6$ intersection points with D_f to s as indicated in Figure 2.5.1.

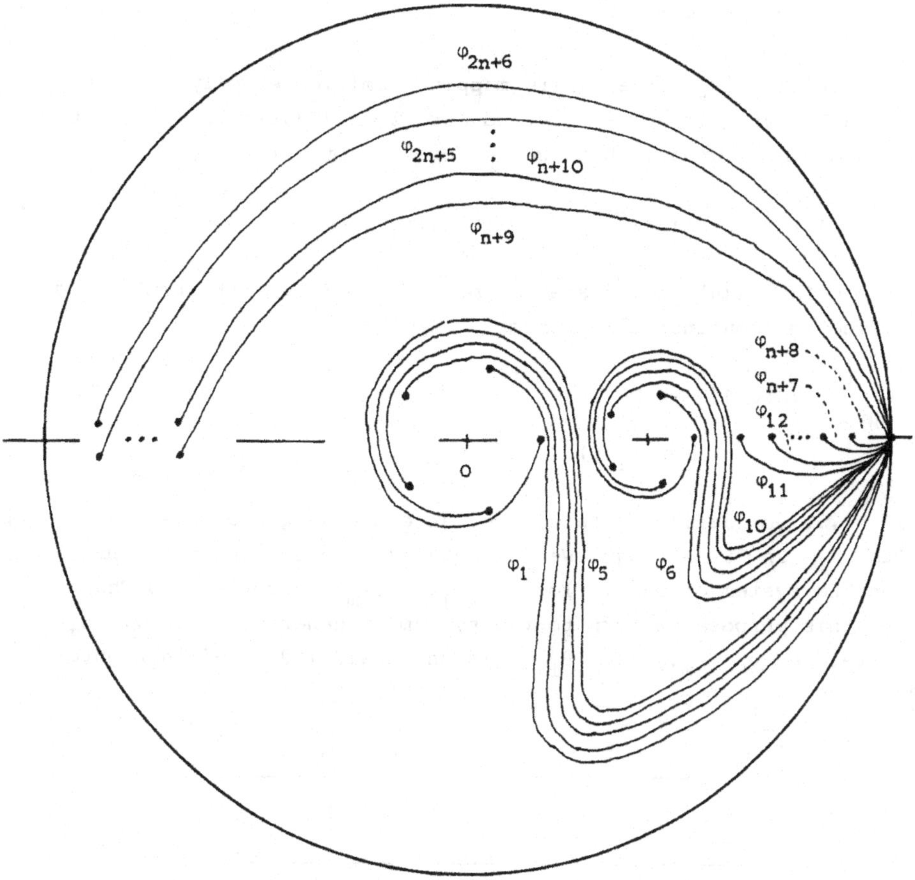

φ_{2n+6}
φ_{2n+5}
φ_{n+10}
φ_{n+9}
φ_{n+8}
φ_{n+7}
φ_{12}
0
φ_{11}
φ_{10}
φ_1
φ_5
φ_6

Figure 2.5.1

Let $(\omega_1,\ldots,\omega_{2n+6})$ be the system of simple loops corresponding to the system of paths $(\varphi_1,\ldots,\varphi_{2n+6})$. By van Kampen's theorem (cf. [Brieskorn$_2$, 2.1.5],[Zariski$_1$]), $(\omega_1,\ldots,\omega_{2n+6})$ define a system of generators of $\pi_1(P-(P\cap D_f),s)$. By this theorem, one gets the corresponding relations as follows. We consider the projection

$$\pi : T\times \mathbb{D} \longrightarrow T.$$

The restriction $\pi|_{D_f}$ is a finite mapping ramified exactly above the origin $0\in T$. Let $T^* = T-\{0\}$, and let $\lambda\in\pi_1(T^*,t)$ be a generator of the fundamental group, which we represent in the form

$$\lambda(\theta) = \eta_1\exp(2\pi\sqrt{-1}\theta), \quad \theta\in[0\ 1].$$

Above the point $\lambda(\theta)$ there are m points $(\lambda(\theta),v_i(\theta))$, $i=1,\ldots,m$ of D_f. The m continuous mappings

$$v_i : [0,1] \longrightarrow \mathbb{D}\ ,$$
$$\theta \longmapsto v_i(\theta)$$

for $i=1,\ldots,m$ define a braid, i.e. an element ζ_f of the braid group Z_m with $m=2n+6$ strings. We now consider the action of the braid group on the system of simple loops $(\omega_1,\ldots,\omega_m)$ according to Chapter 1.5. For this purpose we turn to the system of generators $\beta_2,\ldots,\beta_m$ of the braid group Z_m. Here β_{i+1} is the braid indicated in Figure 2.5.2.

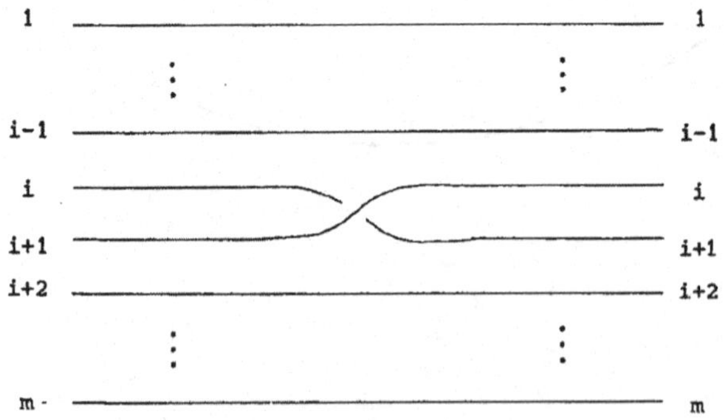

Figure 2.5.2.

The action of β_{i+1} is the same action as described in Chapter 1.5:

$$\beta_{i+1}(\omega_1,\ldots,\omega_m) = (\omega_1,\ldots,\omega_{i-1},\omega_{i+1},\omega_{i+1}\ \omega_i\omega_{i+1}^{-1},\omega_{i+2},\ldots,\omega_m).$$

Van Kampen's theorem says that the relations in $\pi_1(P-(P\cap D_f),s)$ are given by

$$\zeta_f(\omega_i) = \omega_i \quad \text{for} \quad i=1,\ldots,m.$$

We want to compute these relations. We first consider the braid ζ_1 corresponding to D_1 given by $v_1,\ldots,v_5$. Then

$$\zeta_1 = (\beta_2\beta_3\beta_4\beta_5)^6 \quad .$$

Here the braid is written in such a way that one has to apply the elements to the systems of loops from right to left. This yields

$$\zeta_1(\omega_i) = \tau_1\omega_{i+1}\tau_1^{-1} \quad \text{for} \quad i = 1,\ldots,4,$$

$$\zeta_1(\omega_5) = \tau_1^2\omega_1\tau_1^{-2} \quad ,$$

where $\tau_1 = \omega_5\omega_4\omega_3\omega_2\omega_1$. One obtains analogous equations for the braid ζ_2 corresponding to D_2.

For the braid ζ_j corresponding to D_j for $j \geq 3$, which is given by v_{2j+5}, v_{2j+6}, one can compute

$$\zeta_j = \beta_{2j+6}^3 \quad ,$$

$$\zeta_j(\omega_{2j+5}) = \tau_j\omega_{2j+6}\tau_j^{-1} \quad ,$$

$$\zeta_j(\omega_{2j+6}) = \tau_j\omega_{2j+6}\omega_{2j+5}\omega_{2j+6}^{-1}\tau_j^{-1} \quad ,$$

where $\tau_j = \omega_{2j+6}\omega_{2j+5}$.

From these equations, we can derive the action of the whole braid ζ_f by means of the following lemma

Lemma 2.5.1. Let ζ' be a braid with p strings, ζ'' be a braid with q strings, and let ζ be the braid composed of the braids ζ' and ζ'' by linking them once according to Figure 2.5.3.

Then $\zeta = \zeta'\zeta''\zeta_0$, where

$$\zeta_0 = \prod_{j=0}^{q-1} \left(\prod_{i=0}^{p-1} \beta_{p+1+j-i}\right) \prod_{j=0}^{q-1} \left(\prod_{i=0}^{p-1} \beta_{q+1+i-j}\right) .$$

The action of ζ_0 on a strongly distinguished system of loops $(\omega_1,\ldots,\omega_{p+q})$ as above is given by

$$\zeta_0(\omega_\ell) = \begin{cases} \tau''\omega_\ell(\tau'')^{-1} & \text{for } 1 \leq \ell \leq p , \\ \tau(\tau'')^{-1}\omega_\ell\tau''\tau^{-1} & \text{for } p+1 \leq \ell \leq p+q , \end{cases}$$

where $\tau' = \omega_p\omega_{p-1}\ldots\omega_1$, $\tau'' = \omega_{p+q}\omega_{p+q-1}\ldots\omega_{p+1}$ and $\tau = \tau''\tau'$.

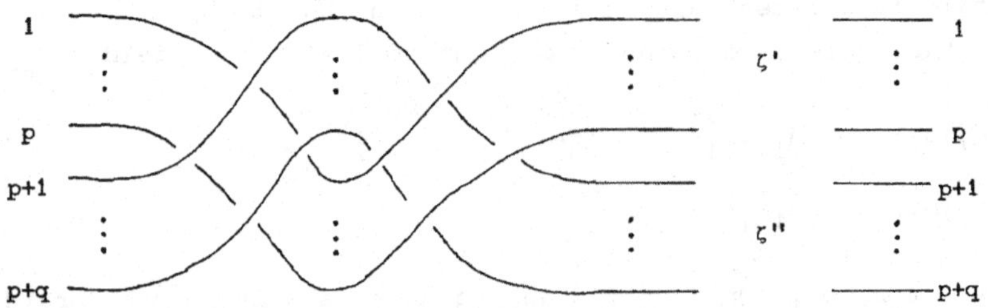

Figure 2.5.3

Proof. The description of ζ_0 is easily derived from Figure 2.5.3. Now let first $1 \leq \ell \leq p$. Then

$$\begin{aligned} \zeta_0(\omega_\ell) &= \prod_{j=0}^{q-1}\left(\prod_{i=0}^{p-1} \beta_{p+1+j-i}\right)(\omega_{\ell+q}) \\ &= \prod_{j=0}^{q-2}\left(\prod_{i=0}^{p-1} \beta_{p+1+j-i}\right)(\omega_{p+q}\omega_{\ell+q-1}\omega_{p+q}^{-1}) \\ &= \omega_{p+q}\ldots\omega_{p+2}\omega_{p+1}\omega_\ell\omega_{p+1}^{-1}\omega_{p+2}^{-1}\ldots\omega_{p+q}^{-1} \\ &= \tau''\omega_\ell(\tau'')^{-1} . \end{aligned}$$

For $p+1 \leq \ell \leq p+q$ one has

$$\zeta_o(\omega_\ell) = \prod_{j=0}^{q-1}\left(\prod_{i=0}^{p-1}\beta_{p+1+j-i}\right)\prod_{j=0}^{q-1-\ell+p}\left(\prod_{i=0}^{p-1}\beta_{q+1+i-j}\right)(\omega_\ell\cdots\omega_{\ell-p+1}\omega_{\ell-p}\omega_{\ell-p+1}^{-1}\cdots\omega_\ell^{-1})$$

$$= \prod_{j=0}^{q-1}\left(\prod_{i=0}^{p-1}\beta_{p+1+j-i}\right)(\omega_{p+q}\cdots\omega_{q+1}\omega_{\ell-p}\omega_{q+1}^{-1}\cdots\omega_{p+q}^{-1})$$

$$= \prod_{j=0}^{\ell-p-2}\left(\prod_{i=0}^{p-1}\beta_{p+1+j-i}\right)(\omega_{p+q}\cdots\omega_{\ell-p}\omega_{\ell+p}^{-1}\cdots\omega_{p+q}^{-1}\omega_\ell\omega_{p+q}\cdots\omega_{\ell+p}\omega_{\ell-p}^{-1}\cdots\omega_{p+q}^{-1})$$

$$= \tau(\tau'')^{-1}\omega_\ell\tau''\tau^{-1} \quad .$$

This proves Lemma 2.5.1.

Combining these results, one finally gets

<u>Proposition 2.5.2.</u> <u>The fundamental group</u> $\pi_1(P - (P \cap D_f), s)$ <u>is the group with generators</u> $(\omega_1, \ldots, \omega_{2n+6})$ <u>and relations</u>

$$\tilde{\tau}\omega_{i+1}\tilde{\tau}^{-1} = \omega_i \quad \underline{\text{for}} \quad 1 \le i \le 4\ , \quad 6 \le i \le 9\ , \quad i = 2j+9\ ,$$

$$\tilde{\tau}\omega_5\omega_4\omega_3\omega_2\omega_1\omega_2^{-1}\omega_3^{-1}\omega_4^{-1}\omega_5^{-1}\tilde{\tau}^{-1} = \omega_5\ ,$$

$$\tilde{\tau}\omega_{10}\omega_9\omega_8\omega_7\omega_6\omega_7^{-1}\omega_8^{-1}\omega_9^{-1}\omega_{10}^{-1}\tilde{\tau}^{-1} = \omega_{10}\ ,$$

$$\tilde{\tau}\omega_{2j+10}\omega_{2j+9}\omega_{2j+10}^{-1}\tilde{\tau}^{-1} = \omega_{2j+10}\ ,$$

<u>where</u> $\tilde{\tau} = \omega_{2n+6}\omega_{2n+5}\cdots\omega_2\omega_1$ <u>and</u> $j = 1,\ldots,n-2$.

Now we consider the strongly distinguished system of generators $(\delta_1,\ldots,\delta_m)$ of vanishing cycles corresponding to the system of paths $(\varphi_1,\ldots,\varphi_m)$, the corresponding Picard-Lefschetz transformations $(\gamma_1,\ldots,\gamma_m)$, and the monodromy operator $c = \gamma_1\gamma_2 \cdots \gamma_m$, $m = 2n+6$. Then we can derive from Proposition 2.5.2:

<u>Corollary 2.5.3.</u> <u>The action of the monodromy operator</u> c <u>on the vanishing cycles</u> δ_i <u>is given by the following equations</u>

$$c(\delta_i) = \pm\delta_{i+1} \quad \underline{\text{for}} \quad 1 \le i \le 4\ , \quad 6 \le i \le 9\ , \quad i = 2j+9\ ,$$

$$c(\delta_5) = \pm\gamma_5\gamma_4\gamma_3\gamma_2(\delta_1)\ ,$$

$$c(\delta_{10}) = \pm\gamma_{10}\gamma_9\gamma_8\gamma_7(\delta_6)\ ,$$

$$c(\delta_{2j+10}) = \pm\gamma_{2j+10}(\delta_{2j+9})\ ,$$

where $j = 1,\ldots,n-2$.

Proof. We only derive the last equation as an example:

$$\tilde{\tau}\omega_{2j+10}\omega_{2j+9}\omega_{2j+10}^{-1}\tilde{\tau}^{-1} = \omega_{2j+10}$$

$$\Rightarrow \quad \gamma_{2j+10}\cdots\gamma_2\gamma_1\gamma_{2j+10}\gamma_{2j+9}\gamma_{2j+10}^{-1}\gamma_1^{-1}\gamma_2^{-1}\cdots\gamma_{2j+10}^{-1} = \gamma_{2j+10}$$

$$\Rightarrow \quad \gamma_{2j+10}\gamma_{2j+9}\gamma_{2j+10}^{-1} = c\gamma_{2j+10}c^{-1} \text{ (since } \gamma_i^2 = \mathrm{id})$$

$$\Rightarrow \quad s^{(n)}_{\delta_{2j+10}}\, s^{(n)}_{\delta_{2j+9}}\, s^{(n)}_{\delta_{2j+10}} = c\, s^{(n)}_{\delta_{2j+10}}\, c^{-1}$$

$$\Rightarrow \quad s^{(n)}_{\gamma_{2j+10}(\delta_{2j+9})} = s^{(n)}_{c(\delta_{2j+10})} .$$

This implies $c(\delta_{2j+10}) = \pm\gamma_{2j+10}(\delta_{2j+9})$, since e.g. $\langle c(\delta_{2j+10}), c(\delta_{2j+10})\rangle \neq 0$. This proves the corollary.

Since $\mu' = 1$ in our example, there is one relation

$$\sum_{i=1}^{2n+6} r_i\delta_i = 0, \quad r_i \in \mathbb{Z} ,$$

between the vanishing cycles δ_i. We set

$$r = \sum_{i=1}^{2n+6} r_i\hat{\delta}_i ,$$

where $\hat{\delta}_i$ is the thimble corresponding to δ_i. We assume in addition that the vanishing cycles δ_i are oriented in such a way that for $1 \leq i \leq 4$, $6 \leq i \leq 9$ and $i = 2j+9$, $j = 1,\ldots,n-2$, one has

$$c(\delta_i) = \delta_{i+1} .$$

Then we can derive the following equations for the relative monodromy operator $\hat{c}$ from Corollary 2.5.3:

Corollary 2.5.4. The relative monodromy operator $\hat{c}$ can be described with respect to the strongly distinguished basis $(\hat{\delta}_1,\ldots,\hat{\delta}_{2n+6})$ of thimbles as follows:

$$\hat{c}(\hat{\delta}_i) = \hat{\delta}_{i+1} + b_i r \qquad \underline{\text{for}}\ 1 \le i \le 4\ ,\ 6 \le i \le 9\ ,\ i = 2j+9\ ,$$

$$\hat{c}(\hat{\delta}_5) = \varepsilon_5 \hat{\gamma}_5 \hat{\gamma}_4 \hat{\gamma}_3 \hat{\gamma}_2 (\hat{\delta}_1) + b_5 r\ ,$$

$$\hat{c}(\hat{\delta}_{10}) = \varepsilon_{10} \hat{\gamma}_{10} \hat{\gamma}_9 \hat{\gamma}_8 \hat{\gamma}_7 (\hat{\delta}_6) + b_{10} r\ ,$$

$$\hat{c}(\hat{\delta}_{2j+10}) = \varepsilon_{2j+10} \hat{\gamma}_{2j+10} (\hat{\delta}_{2j+9}) + b_{2j+10} r\ ,$$

<u>for</u> $b_i \in \mathbb{Z}$, $\varepsilon_i = \pm 1$, $j = 1, \ldots, n-2$.

We now want to determine the intersection matrix

$$A = ((a_{ij})) = ((\langle \hat{\delta}_i, \hat{\delta}_j \rangle))$$

of the thimbles $\hat{\delta}_i$. We use the relation of Proposition 1.6.3 for the matrix $\hat{C}$ of the relative monodromy operator $\hat{c}$ with respect to the basis $(\hat{\delta}_1, \ldots, \hat{\delta}_{2n+6})$:

$$\hat{C} = -V^{-1} V^t\ ,$$

where $A = V + V^t$ and V is an upper triangular matrix with $\varepsilon = (-1)^{n/2}$ on the diagonal (n even in our case). This yields the matrix equation

$$V^t = -V\hat{C} \qquad (*)$$

By (1.6.2) one has

$$\hat{c}(r) = r \qquad (**)$$

We use the equations (*) and (**) to determine recursively the intersection numbers a_{ij} and the coefficients r_i of the relation between the δ_i by means of Corollary 2.5.4. Setting

$$b_0 = 1 - \sum_{i=1}^{m} b_i r_i\ ,$$

(**) implies, by comparing coefficients, that

$$\varepsilon_m r_m = b_0 r_{m-1}\ ,$$

$$-\varepsilon \varepsilon_m a_{m-1,m} r_m + r_{m-1} = b_0 r_m\ .$$

If $b_0 = 0$ then $r_{m-1} = r_m = 0$, and then (*) implies $a_{i,m-1} = a_{i,m} = 0$ for $i = 1,\dots,m-2$. But this is impossible, since the Dynkin diagram corresponding to A is connected (by Remark 1.5.2). Therefore $b_0 \neq 0$, and hence

$$r_{m-1} = \varepsilon_m r_m / b_0 \quad ,$$

$$a_{m-1,m} = -\varepsilon\varepsilon_m b_0 + (\varepsilon/b_0) .$$

We now use equation (*). We denote by (i,j) the equation obtained from (*) by comparing the entry in row i and column j on the right-hand side with the corresponding entry on the left-hand side of the matrix equation (*).

Step (1). We first consider the following equations

$$(m,m) \qquad -1 = b_o - (\varepsilon_m/b_o) + b_m r_m$$

$$(m-1,m) \qquad 0 = \varepsilon_m + \varepsilon_m(b_m r_m/b_o) + \varepsilon_m b_o - (1/b_o)$$

$$(m,m-1) \qquad -\varepsilon_m b_o + (1/b_o) = -1 - b_{m-1} r_m$$

These equations imply

$$b_0 = \varepsilon_m^2 = 1, \quad b_{m-1} = b_m \;, \quad b_m r_m = \varepsilon_m - 2.$$

If $\varepsilon_m = -1$ then $b_m r_m = -3$. We set

$$b_{m-1} = b_m := \pm 1.$$

This is allowed if we admit from now on that $b_i \in \mathbf{Z}$ or $b_i \in (1/3)\mathbf{Z}$ for $i = 1,\dots,m-2$.

Step (q), $q = 2,\dots,n-2$. Let $1 \leq p \leq q$. We assume that the following equations are valid (for $p = 1$ these equations are true according to step (1)).

(a) for $i = m - 2p + 1 , m - 2p + 2 ; \; i < j \le m$:

$$(j,i) \begin{cases} \varepsilon a_{ij} = (2 - \varepsilon_m) b_i b_j & (i,j) \ne (m-2p+1, m-2p+2) \\ \varepsilon a_{ij} = -1 + (2 - \varepsilon_m) b_i b_j & (i,j) = (m-2p+1, m-2p+2) \end{cases}$$

(b) for $i = m - 2p + 1 , m - 2p + 2$:

$$(i,m) \quad r_i = -(-\varepsilon_m)^{p-1}(2 - \varepsilon_m) b_i - \sum_{\ell=i+1}^{m-2p+2} a_{i\ell} r_\ell$$

(c) $(m - 2p + 1 , m - 2p + 1) \quad 0 = b_{m-2p+1}(b_{m-2p+1} - b_{m-2p+2})$

(d) $a_{m-2p+1,m-2p+2} = -\varepsilon\varepsilon_{m-2p+2} + \varepsilon$

(This equation follows from (**) as above. Note that $b_0 = 1$).

(e) $\varepsilon_{m-2p+2} = \varepsilon_m \; , \; b_{m-2p+1} = b_{m-2p+2} = \pm 1$

From the validity of the equations (a) - (e) for $p < q$ and from the equations (*) and (**) follows that the equations (a) - (d) are valid for $p = q$. Equation (d) for $p = q$ and equations $(m - 2q + 1, m - 2q + 2)$ and $(m - 2q + 1, m - 2q + 1)$ then imply equation (e) for $p = q$.

<u>Step (n-1).</u> In this step we consider the following equations. Let $p = 5$.

(a) For $p+1 \le i \le p+5 \; , \; i < j \le m$:

$$(j,i) \begin{cases} \varepsilon a_{ij} = (2 - \varepsilon_m) b_i b_j & \text{for } j \ne i+1 , \\ \varepsilon a_{ij} = -1 + (2 - \varepsilon_m) b_i b_j & \text{otherwise} , \end{cases}$$

(b) for $p+1 \le i \le p+5$:

$$(i,m) \quad r_i = -(2 - \varepsilon_m) b_i - \sum_{\ell=i+1}^{p+5} a_{i\ell} r_\ell ,$$

(c) for $p+1 \le i \le j \le p+5 \; , \; j \ne p+5$:

$$(i,j) \quad 0 = b_i (b_{j+1} - b_j) .$$

(d) Equation (**) with $b_0 = 1$ implies:

$$\varepsilon_{p+5} r_{p+5} = r_{p+1} ,$$

$$-\varepsilon\varepsilon_{p+5} a_{p+1,p+2} r_{p+5} + r_{p+1} = r_{p+2} ,$$

$$r_{p+1} = 0 \Rightarrow r_{p+1} = r_{p+2} = r_{p+3} = r_{p+4} = r_{p+5} = 0 .$$

We now conclude as follows:

Suppose $b_{p+1} = 0$. Then

$$a_{p+1,p+2} = -1 \qquad \text{(by (p+2,p+1))}$$

$$\Rightarrow \quad r_{p+2} = 0 \qquad \text{(by (d))}$$

$$\Rightarrow \quad r_{p+1} = 0 \qquad \text{(by (p+1,m))}$$

$$\Rightarrow \quad r_{p+1} = r_{p+2} = r_{p+3} = r_{p+4} = r_{p+5} = 0 \qquad \text{(by (d))}$$

$$\Rightarrow \quad b_{p+1} = b_{p+2} = b_{p+3} = b_{p+4} = b_{p+5} = 0 \qquad \text{(by (b))} .$$

Then (a) implies that the Dynkin diagram corresponding to $(\hat{\delta}_1, \ldots, \hat{\delta}_m)$ is not connected. Hence $b_{p+1} \neq 0$. But then (c) implies

$$b_{p+1} = b_{p+2} = b_{p+3} = b_{p+4} = b_{p+5} \neq 0 .$$

Then (a), (b) and (d) yield the following equation for b_{p+5} :

$$\varepsilon_{p+5} = 5 - 20(2-\varepsilon_m)b_{p+5}^2 + 21(2-\varepsilon_m)^2 b_{p+5}^4 - 8(2-\varepsilon_m)^3 b_{p+5}^6 + (2-\varepsilon_m)^4 b_{p+5}^8 .$$

This equation has exactly the solutions

$$b_{p+5}^2 = 1, \quad \varepsilon_{p+5} = \varepsilon_m .$$

<u>Step (n).</u> Using step (n-1), the equations (*) and (**) imply that the equations (a) - (d) of step (n - 1) are also valid for $p = 0$, and we can conclude as in step (n - 1).

<u>Result:</u> Up to changes of orientations of the thimbles $\hat{\delta}_i$, the equations (*) and (**) have the two solutions $(V^{(+1)}, r^{(+1)})$ for $\varepsilon_m = +1$ and $(V^{(-1)}, r^{(-1)})$ for $\varepsilon_m = -1$. Here $V^{(\pm 1)} = ((v_{ij}^{(\pm 1)}))$, and for $i < j$

$$v_{ij}^{(\pm 1)} = a_{ij}^{(\pm 1)} = \begin{cases} (-1)^{n/2}(1 \mp 1) & \text{for } j = i+1 \text{ and } i \neq 5,10;\ 10+2p; \\ & p = 1,\dots,n-3 \ , \\ (-1)^{n/2}(2 \mp 1) & \text{otherwise} \ , \end{cases}$$

and

$$r_i^{(+1)} = \begin{cases} 1 & \text{for } i = 1,2;6,7;7+4q,8+4q; \quad q = 1,\dots,(n-2)/2 \ ; \\ 0 & \text{for } i = 3;8; \\ -1 & \text{for } i = 4,5;9,10;9+4q,10+4q; \quad q = 1,\dots,(n-2)/2 \ ; \end{cases}$$

$$r_i^{(-1)} = \begin{cases} 3 & \text{for } i = 1,4;6,9;9+2p; \quad p = 1,\dots,n-2 \ ; \\ 0 & \text{for } i = 3;8; \\ -3 & \text{for } i = 2,5;7,10;10+2p; \quad p = 1,\dots,n-2 \ . \end{cases}$$

We show that the solution $(V^{(-1)}, r^{(-1)})$ is not possible. For this purpose we make the following basis substitution:

$$e_1 = \hat{\delta}_1 - \hat{\delta}_2 + \hat{\delta}_4 - \hat{\delta}_5 \ , \quad e_2 = \hat{\delta}_6 - \hat{\delta}_7 + \hat{\delta}_9 - \hat{\delta}_{10} \ ;$$

$$e_{2+p} = \hat{\delta}_{2p+9} - \hat{\delta}_{2p+10} \ , \quad p = 1,\dots,n-2 \ ;$$

$$e_{n+1} = \hat{\delta}_1 - \hat{\delta}_2 \ , \quad e_{n+2} = \hat{\delta}_3 - \hat{\delta}_2 \ ; \quad e_{n+3} = \hat{\delta}_6 - \hat{\delta}_7 \ , \quad e_{n+4} = \hat{\delta}_8 - \hat{\delta}_7 \ ;$$

$$e_{n+5} = -\hat{\delta}_1 + 2\hat{\delta}_2 - \hat{\delta}_4 \ , \quad e_{n+6} = \hat{\delta}_2 \ , \quad e_{n+7} = -\hat{\delta}_6 + 2\hat{\delta}_7 - \hat{\delta}_9 \ ,$$

$$e_{n+8} = \hat{\delta}_7 \ ; \quad e_{n+8+p} = \hat{\delta}_{10+2p} \ , \quad p = 1,\dots,n-2 \ .$$

Under the assumption that the intersection matrix corresponding to $(\hat{\delta}_1,\dots,\hat{\delta}_{2n+6})$ is the matrix $A^{(-1)}$, we can derive that

$$\ker(\hat{H}) = \sum_{i=1}^{n} \mathbb{Z} e_i \ , \quad \sum_{i=n+1}^{n+4} \mathbb{Z} e_i = U \perp U \ , \quad \sum_{i=n+5}^{2n+6} \mathbb{Z} e_i =: L \ ,$$

$$\hat{H} = \ker(\hat{H}) \perp U \perp U \perp L \ ,$$

where U is a unimodular hyperbolic plane and L an indefinite lattice of rank $n+2$. This yields the following upper bound for the number μ_ε of eigenvalues with sign ε, $\varepsilon = (-1)^{n/2}$, of the quadratic form on $\hat{H}$, respectively on H:

$$\mu_\varepsilon \leq n + 3 \ .$$

But it follows from Chapter 3 that T^n_{2233} deforms to the singularity T^n_{2223} for which $\mu_\varepsilon = n+4$. By [Looijenga$_3$, (7.13)] we get a contradiction.

We have therefore proved

Proposition 2.5.5. *The n-dimensional singularity T^n_{2233} with n even has a strongly distinguished basis $(\hat{\delta}_1,\ldots,\hat{\delta}_{2n+6})$ of thimbles with the following intersection numbers*

$$\langle\hat{\delta}_i,\hat{\delta}_j\rangle = \begin{cases} 0 & \text{for } j = i+1 \text{ and } i \neq 5;10;2p+10; \quad p = 1,\ldots,n-2, \\ (-1)^{n/2} & \text{for } i < j \quad \text{otherwise.} \end{cases}$$

There is the following relation between the corresponding vanishing cycles $(\delta_1,\ldots,\delta_{2n+6})$:

$$\delta_1 + \delta_2 - \delta_4 - \delta_5 + \delta_6 + \delta_7 - \delta_9 - \delta_{10} + \sum_{j=1}^{n-2} (-1)^{j+1}(\delta_{2j+9} + \delta_{2j+10}) = 0 \ .$$

We now transform this basis by means of the braid group action to a basis with a Dynkin diagram which is an extension of the graph of Figure 2.3.7. By the transformations

$$\begin{array}{llll} \beta_{2n+3}, & \beta_{2n+4}, & \beta_{2n+5}, & \beta_{2n+6} \ ; \\ \beta_{2n+2}, & \beta_{2n+3}, & \beta_{2n+4}, & \beta_{2n+5} \ ; \\ \vdots & \vdots & \vdots & \vdots \\ \beta_7, & \beta_8, & \beta_9, & \beta_{10} \end{array}$$

the basis $(\hat{\delta}_1,\dots,\hat{\delta}_{2n+6})$ passes into a basis $(\hat{\delta}'_1,\dots,\hat{\delta}'_{2n+6})$ with

$$\langle\hat{\delta}'_i,\hat{\delta}'_j\rangle = \begin{cases} 0 & \text{for } j = i+1 \text{ and } i \neq 5,7,\dots,2n+1, \\ (-1)^{n/2} & \text{for } i < j \text{ otherwise.} \end{cases}$$

By the transformations

$$\alpha_1,\alpha_2,\alpha_3,\alpha_1,\alpha_2,\alpha_3,\alpha_2,\alpha_3,\varkappa_4 ;$$

$$\beta_{2n+6},\beta_{2n+5},\beta_{2n+4},\beta_{2n+6},\beta_{2n+5},\beta_{2n+4},\beta_{2n+5},\beta_{2n+4},\varkappa_{2n+3} ;$$

$$\beta_6,\beta_5;\beta_8,\beta_7,\beta_6,\beta_5;\dots;\beta_{2n},\beta_{2n-1},\dots,\beta_5 ;$$

$$\beta_{n+2},\beta_{n+3};\beta_{n+1},\beta_{n+2};\dots;\beta_5,\beta_6 ;$$

$$\varkappa_1,\varkappa_2,\varkappa_3,\varkappa_4,\varkappa_5;\varkappa_{n+2},\varkappa_n,\dots,\varkappa_6;\varkappa_{n+5},\varkappa_{n+7},\dots,\varkappa_{2n+1},$$

this basis passes into the basis $(\hat{\delta}''_1,\dots,\hat{\delta}''_{2n+6})$ with the Dynkin diagram shown in Figure 2.5.4 (cf. Figure 2.3.7).

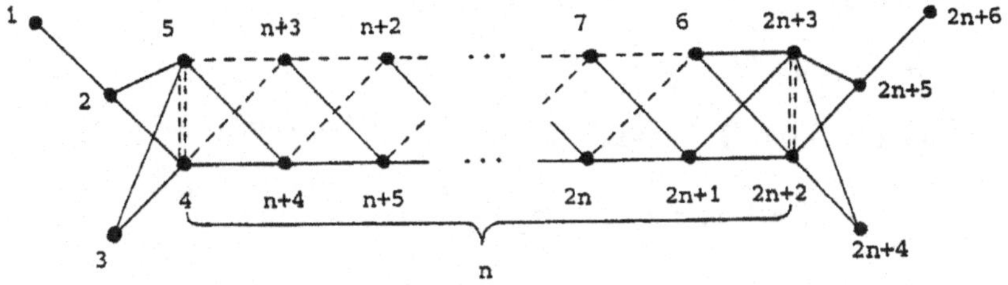

Figure 2.5.4

Applying the transformations

$$\alpha_3,\alpha_2;\beta_6,\beta_7,\dots,\beta_{n+4};\alpha_2,\beta_5,\beta_6,\dots,\beta_{n+4} ;$$

$$\beta_{n+3},\beta_{n+2};\alpha_2,\dots,\alpha_n;\beta_n,\dots,\beta_4;\alpha_{n+1},\alpha_{n+3} ;$$

$$\varkappa_1,\varkappa_2,\varkappa_3,\varkappa_{n+2},\varkappa_{n+3}$$

to this basis yields a strongly distinguished basis $(\lambda_1,\dots,\lambda_{2n+6})$ of thimbles with the Dynkin diagram shown in Figure 2.5.5.

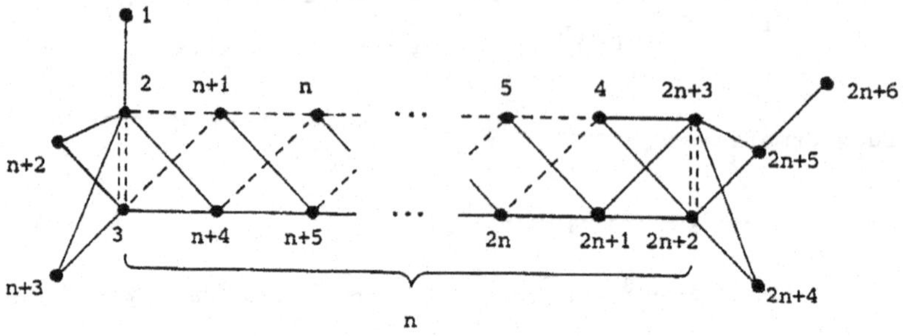

Figure 2.5.5

This is a special basis in the sense of Definition 5.4.1. With the notations introduced in Chapter 4.1 one gets

$$\hat{H} = Q^n_{2223} \perp U \perp \ker \hat{H}\ , \quad \dim(\ker \hat{H}) = n-1\ ,$$

where the lattice Q^n_{2223} is spanned by the vectors $\lambda_3;\lambda_{n+2},\lambda_{n+3},\dots,\lambda_{2n+1};\lambda_{2n+3},\lambda_{2n+4},\lambda_{2n+5},\lambda_{2n+6}$ and the radical $\ker \hat{H}$ by the vectors

$$\lambda_{2n+2} - \lambda_{2n+3}\ ;$$

$$\lambda_{n+4+i} + \lambda_{n+3+i} + \lambda_{n+2-i} - \lambda_{n+1-i}\ , \quad i = 1,\dots,n-3\ ;$$

$$\lambda_{n+4} + \lambda_{n+3} + \lambda_{n+2} - \lambda_{n+1} + 2\lambda_3\ .$$

Note that the lattice Q^n_{2223} (for its definition see Remark 4.1.4) can be decomposed as follows:

$$Q^n_{2223} = D_{n+3} \perp U\ .$$

3. DYNKIN DIAGRAMS FOR SOME SPECIAL SINGULARITIES

3.1. On the classification of isolated singularities of complete intersections

In this chapter we shall apply the methods of Chapter 2 in order to compute Dynkin diagrams of some special singularities. All these singularities will be isolated complete intersection singularities given by map-germs

$$f : (\mathbb{C}^{n+2},0) \longrightarrow (\mathbb{C}^2,0) \quad .$$

For the classification of such singularities, the 2-jet plays a central rôle.

We assume throughout that $df(0) = 0$ (otherwise one can take n smaller in order to satisfy this condition). Let $j^2f = (q_1,q_2)$ be the 2-jet of f, i.e. q_1,q_2 are the quadratic parts of the component functions f_1,f_2 of f. Then

$$\xi_1 q_1 + \xi_2 q_2$$

defines a pencil of quadrics. Such pencils were classified by Weierstraß [Weierstraß$_1$] (in the regular case) and Kronecker [Kronecker$_1$] (in general). The pencil is called <u>regular</u> (non-degenerate) if some quadric of the pencil is non-degenerate. We shall only consider this case in what follows. Without loss of generality, we can then assume that q_1 is non-degenerate. This means that f_1 is a non-degenerate function at the origin, thus has an ordinary double point at the origin. Let Q_i be the matrix of q_i, i.e.

$$q_i(z) = z^tQ_iz \quad .$$

Then the Jordan normal form of $Q_1^{-1}Q_2$ is an invariant of the pencil. We indicate the Jordan normal form in the usual notation by the <u>Segre symbol</u> (cf. [Wall$_4$]). This symbol consists of the list of the sizes of the Jordan blocks with those corresponding to a given eigenvalue enclosed in parentheses. We can also include regular pencils with q_1 degenerate, if we allow ∞ as an eigenvalue. The eigenvalues therefore belong to $\mathbb{P}^1(\mathbb{C})$, and their cross-ratios are invariants of the pencil.

The unimodal isolated complete intersection singularities were classified by C.T.C. Wall [Wall_3]. He considers the classification of map-germs modulo K-equivalence (or contact-equivalence). Here K is the group of diffeomorphisms of $(\mathbb{C}^{n+2} \times \mathbb{C}^2, (0,0))$ of the form $h(z,y) = (h_1(z), h_2(z,y))$, which acts on the map-germs $f : (\mathbb{C}^{n+2},0) \longrightarrow (\mathbb{C}^2,0)$ as follows: For $h = (h_1,h_2) \in K$ one has $(h.f)(z) = h_2(h_1(z), f(h_1(z)))$. Two map-germs f and g are K-equivalent if and only if the germs of the corresponding zero sets $f^{-1}(0)$ and $g^{-1}(0)$ are analytically isomorphic.

An important tool for the classification is a splitting theorem of C.T.C. Wall [Wall_2] which has some analogy with Thom's splitting lemma for the hypersurface case. In order to state this theorem, we consider the subgroup K' of K consisting of those diffeomorphisms $h = (h_1,h_2)$ of $(\mathbb{C}^{n+2} \times \mathbb{C}^2, (0,0))$ as above for which the differential dh' of h' defined by $h'(z,y) = (z, h_2(z,y))$ at $(0,0)$ is equal to the identity. We call f good, if f defines an isolated singularity (i.e. 0 is an isolated singular point of $f^{-1}(0)$) and if $j^2 f$ is regular. For $g : (\mathbb{C}^s,0) \longrightarrow (\mathbb{C}^2,0)$ and $g' : (\mathbb{C}^t,0) \longrightarrow (\mathbb{C}^2,0)$ we define their sum

$$g \oplus g' : (\mathbb{C}^{s+t},0) \longrightarrow (\mathbb{C}^2,0)$$

by $(g \oplus g')(u,v) = g(u) + g'(v)$ for $u \in \mathbb{C}^s$ and $v \in \mathbb{C}^t$.

Theorem 3.1.1 (Splitting theorem of C.T.C. Wall). *Let $f:(\mathbb{C}^{n+2},0) \to (\mathbb{C}^2,0)$ be a good map-germ. Given any partition of the set of eigenvalues of $j^2 f$ into two subsets Λ_1 and Λ_2, there exists a corresponding K'-equivalence*

$$f \sim g \oplus g' ,$$

where g and g' are good and $j^2 g$ has the eigenvalues of Λ_1 and $j^2 g'$ has the eigenvalues of Λ_2. Moreover, the K'-equivalence classes of g and g' are determined by that of f. Conversely, if g and g' are good and their 2-jets have no common eigenvalue, then $g \oplus g'$ is good; and the K'-equivalence classes of g and g' determine that of $g \oplus g'$.

By means of this splitting theorem one obtains the following classification of good map-germs f with eigenvalues of $j^2 f$ of multiplicity less than or equal to 2 modulo K'-equivalence.

Case (1). The multiplicities of the eigenvalues are all equal to 1.

Then f is K'-equivalent to an isolated singularity of an intersection of two quadrics, hence to a singularity of type $\tilde{D}_{n+3}$ (see Chapter 2.3).

Case (2). The multiplicities of the eigenvalues are less than or equal to 2, and equal to 2 for at least one and at most two eigenvalues.

In the case of one eigenvalue of multiplicity 2 we can reduce $f = (f_1, f_2)$ to the following form (cf. [Wall_2], [Mather_1]):

$$f_1(z) = z_1^2 + z_2^2 + \ldots + z_n^2 + 2z_{n+1}z_{n+2} \quad ,$$

$$f_2(z) = a_1z_1^2 + a_2z_2^2 + \ldots + a_nz_n^2 + z_{n+1}^q + z_{n+2}^s \quad ,$$

where $a_i \in \mathbb{C}$, $a_i \neq a_j \neq 0$ for $i \neq j$, $1 \leq i,j \leq n$, and $2 \leq q \leq s$, $3 \leq s$. Here $q = 2$ for the Segre symbol $\{1,\ldots,1,2\}$ and $q \neq 2$ for the Segre symbol $\{1,\ldots,1,(1,1)\}$. We denote these singularities by $T^n_{2,q,2,s}$.

In the case when two eigenvalues have multiplicity 2 we can reduce $f = (f_1, f_2)$ to the following form:

$$f_1(z) = z_1^2 + z_2^2 + \ldots + z_{n-2}^2 + z_{n-1}^p + z_n^r + 2z_{n+1}z_{n+2} \quad ,$$

$$f_2(z) = a_1z_1^2 + a_2z_2^2 + \ldots + a_{n-2}z_{n-2}^2 + 2z_{n-1}z_n + z_{n+1}^q + z_{n+2}^s \quad ,$$

where $a_i \in \mathbb{C}$, $a_i \neq a_j \neq 0$ for $i \neq j$, $1 \leq i,j \leq n-2$, and $2 \leq q \leq s$, $2 \leq p \leq r$, $r \leq s$, $3 \leq s$. We denote these singularites by $T^n_{p,q,r,s}$. For $n = 2$ these are just the two-dimensional cusp singularities of embedding dimension 4.

Case (3). At least one eigenvalue has multiplicity greater than 2.

The simplest case with an eigenvalue of multiplicity greater than 2 is the case with Segre symbol $\{1,\ldots,1,3\}$: it follows from the deformation theory of regular pencils of quadrics (see e.g. [Gantmacher_1]) that all the other regular pencils of case (3) deform into a pencil with this Segre symbol. The splitting theorem and the classification of Wall [Wall_4] imply that a map-germ $f = (f_1, f_2)$ with Segre symbol $\{1,\ldots,1,3\}$ can be reduced to the following so-called prenormal form

$$f_1(w;x,y,z) = w_1^2 + \ldots + w_{n-1}^2 + 2xz - y^2 ,$$

$$f_2(w;x,y,z) = a_1w_1^2 + \ldots + a_{n-1}w_{n-1}^2 + 2xy + Q(y,z) .$$

Here $a_i \in \mathbb{C}$, $a_i \neq a_j \neq 0$ for $i \neq j$, $1 \leq i,j \leq n-1$ and

$$Q(y,z) = a(z) + yb(z) + y^2c(z)$$

with $\operatorname{ord} a \geq 3$, $\operatorname{ord} b \geq 2$, $\operatorname{ord} c \geq 1$. In fact $c(z)$ is superfluous and can always be set equal to zero. However, we use this term in order to simplify some equations. For $n = 1$ the map-germ f is equal to $\tilde{f}$ with

$$\tilde{f}(x,y,z) = (2xz - y^2, 2xy + Q(y,z)) .$$

Depending on the choice of $a(z)$, $b(z)$, and $c(z)$, $\tilde{f}$ belongs to different μ-constant strata. Wall calls the corresponding series of singularities with this prenormal form the J-series. For general n the map-germ f can be written as

$$f = (w_1^2, a_1w_1^2) \oplus \ldots \oplus (w_{n-1}^2, a_{n-1}w_{n-1}^2) \oplus \tilde{f} .$$

In the case $n = 2$ Wall denotes the corresponding series by J'. Analogously, we call this series of singularities for general n the $J^{(n-1)}$-series.

The next case with exactly one eigenvalue of multiplicity 3 and otherwise only eigenvalues of multiplicity 1 after the case with Segre symbol $\{1,\ldots,1,3\}$ is the case with Segre symbol $\{1,\ldots,1,(1,2)\}$. A map-germ $f = (f_1,f_2)$ with this Segre symbol can be reduced to the following prenormal form by Theorem 3.1.1 and [Wall_4]:

$$f_1(w;x,y,z) = w_1^2+\ldots+w_{n-1}^2 + 2xz + y^2 ,$$

$$f_2(w;x,y,z) = a_1w_1^2+\ldots+a_{n-1}w_{n-1}^2 + x^2 + Q(y,z).$$

Here again $a_i \in \mathbb{C}$, $a_i \neq a_j \neq 0$ for $i \neq j$, $1 \leq i,j \leq n-1$, and $Q(y,z)$ as above. For $n = 1$ one has $f = \tilde{f}$ with

$$\tilde{f}(x,y,z) = (2xz + y^2, x^2 + Q(y,z)).$$

As above, for $n \geq 2$ the map-germ f can be written as

$$f = (w_1^2, a_1 w_1^2) \oplus \ldots \oplus (w_{n-1}^2, a_{n-1} w_{n-1}^2) \oplus \tilde{f} .$$

For $n = 1$ Wall calls the series of singularities with the above Segre symbol the K-series. By analogy to the case $J^{(n-1)}$, we call this series of singularities for any n the $K^{(n-1)}$-series.

The map-germs $f: (\mathbb{C}^{n+2}, 0) \longrightarrow (\mathbb{C}^2, 0)$ with an isolated singularity and regular 2-jet have been partially classified by Wall for $n = 1,2$ ([Wall_3],[Wall_4]). Wall classifies the K-unimodal map-germs, and he restricts his attention to those Segre symbols for which there exist K-unimodal map-germs in the corresponding class of map-germs. In case (3) one has the following table with the possible Segre symbols and the corresponding notations of the series by Wall:

$n = 1$:

Segre Symbol	{3}	{(1,2)}
Series	J	K

$n = 2$:

Segre Symbol	{1,3}	{1,(1,2)}	{4}	{(1,3)}
Series	J'	K'	L	M

Wall also considers some cases with singular 2-jet. But we restrict ourselves here to the regular case.

We shall compute Dynkin diagrams for the singularities $T^n_{2,q,2,s}$ in Section 3.2 and for the singularities of the $J^{(n-1)}$- and the $K^{(n-1)}$-series in Section 3.3. In Section 3.4 we consider Dynkin diagrams for the simple space curve singularities $(n = 1)$, and finally in Sections 3.5 and 3.6 Dynkin diagrams for singularities of surfaces $(n = 2)$ of complete intersection, given by map-germs belonging to the above series.

3.2. The singularities $T^n_{2,q,2,s}$

In this section we compute Dynkin diagrams of the singularities of type $T^n_{2,q,2,s}$ introduced in the last section by means of the method presented in Chapter 2.2. We give a detailed description of the procedure in this case. This procedure is similar in all cases. We first consider the case

(A) Segre symbol $\{1,\dots,1,(1,1)\}$

For the calculation we use the mapping $f = (f_1, f_2)$ with

$$f_1(z) = z_1^2 + z_2^2 + \dots + z_n^2 + z_{n+1}^2 - z_{n+2}^2 ,$$

$$f_2(z) = a_1 z_1^2 + a_2 z_2^2 + \dots + a_n z_n^2 + b_0(z_{n+1} + z_{n+2})^q + b_0(z_{n+1} - z_{n+2})^s .$$

Here $3 \leq q \leq s$, b_0 is an arbitrary non-real complex number, and the numbers a_i must satisfy the following conditions

$$a_i \in \mathbb{C} , \quad -\frac{1}{4} < \operatorname{Re} a_i < \operatorname{Re} a_j < 0 \quad \text{for } i < j ,$$

$$0 < \operatorname{Im} a_i = \operatorname{Im} a_j \ll 1 \quad \text{for} \quad 1 \leq i,j \leq n ,$$

where the exact condition on $\operatorname{Im} a_i$ will be fixed later. As the linear function ζ we choose the last coordinate function z_{n+2}. Then an easy calculation shows that

$$\Sigma_{z_{n+2}}(f_2) = \bigcup_{i=1}^{2n+2} \Sigma_i ,$$

where

$$\Sigma_{2j-\varkappa} = \{z_i = 0 \text{ for } 1 \leq i \neq j \leq n , \ z_j = g_{2j-\varkappa}(z_{n+2}) , \ z_{n+1} = \tilde{g}_{2j-\varkappa}(z_{n+2})\} ,$$

$$\Sigma_{2n+2-\varkappa} = \{z_i = 0 \text{ for } i = 1,\dots,n , \ z_{n+1} = (-1)^{\varkappa+1} z_{n+2}\} ,$$

for $1 \leq j \leq n$, $\kappa \in \{0,1\}$; $g_{2j-\kappa}(z_{n+2})$ being a power series in z_{n+z} beginning with $(-1)^{\kappa} z_{n+2}$, and $\tilde{g}_{2j-\kappa}(z_{n+2})$ being a power series of order $q-1$. One finds

$$\rho_i = 2 \quad \text{for} \quad i = 1,\dots,2n, \ \rho_{2n+1} = q, \ \rho_{2n+2} = s.$$

The function ζ was chosen in such a way that the 2-jet of $g = (f_1, f_2 + \zeta^2)$ has only eigenvalues of multiplicity 1. By Section 3.1 g defines a singularity of type $\tilde{D}_{n+3}$. For the application of Theorem 2.2.3, we have to determine the intersection matrix of a basis $(e_1,\dots,e_\nu)$ of thimbles of this singularity defined by a strongly distinguished system of paths described in Chapter 2.2. This is a system of paths joining the critical values of the function

$$g_{\varepsilon,t} = (f_2 + (z_{n+2} - \varepsilon)^2)|_{X_t'}$$

to the origin, where $0 < \delta << \varepsilon << 1$, $0 < t \leq \delta$, and X_t' is the Milnor fibre of f_1 above the point t. In order to determine the intersection matrix of a basis of thimbles corresponding to such a system of paths, we compare this system of paths with that of Chapter 2.3 for the function $g_{0,t}$. For this purpose we have to study the asymptotic behaviour of the critical values of $g_{\varepsilon,t}$ for $\varepsilon \longrightarrow 0$. One calculates the following asymptotic approximation formulas for the critical values $\tilde{s}_i^{\varepsilon,t}$ corresponding to the critical points of $g_{\varepsilon,t}$ which tend to 0 for $\varepsilon \longrightarrow 0$ and $t \longrightarrow 0$. In each case we indicate the branches of the polar curve on which the corresponding critical points lie for $t = 0$ in parentheses:

$$\tilde{s}_j^{\varepsilon,t} = a_j t + \frac{a_j}{1 + a_j}\varepsilon^2 + o(\varepsilon^q), \quad j = 1,\dots,n, \quad (\Sigma_{2j-1}, \Sigma_{2j})$$

$$\tilde{s}_{2n+\varkappa}^{\varepsilon,t} = (\varepsilon + (-1)^{\varkappa+1}\sqrt{-t})^2 + o(t^{q/2}), \quad \varkappa = 1,4, \quad \text{(origin)}$$

$$\tilde{s}_{2n+2+\varkappa}^{\varepsilon,t} = b_0((-1)^{\varkappa}\sqrt{t+\varepsilon^2} + \varepsilon)^q + b_0((-1)^{\varkappa}\sqrt{t+\varepsilon^2} - \varepsilon)^s + o(\varepsilon^{q+1}),$$
$$\varkappa = 0,1, \quad (\Sigma_{2n+1+\varkappa})$$

We can now formulate the condition on the imaginary parts of the complex numbers a_j: $\operatorname{Im} a_j$ has to be chosen so small that the following condition is satisfied for all $j = 1,\dots,n$:

$$\operatorname{Re}(\tilde{s}_j^{\varepsilon,t}) = \operatorname{Re}(\tilde{s}_{2n+1}^{\varepsilon,t}) \Rightarrow \operatorname{Im}(\tilde{s}_j^{\varepsilon,t}) < \operatorname{Im}(\tilde{s}_{2n+1}^{\varepsilon,t}) \qquad (3.2.1)$$

Since for $1 \leq i < j \leq n$ one has $-\frac{1}{4} < \operatorname{Re} a_i < \operatorname{Re} a_j < 0$ and $0 < \operatorname{Im} a_i = \operatorname{Im} a_j << 1$ by assumption, one gets

$$\operatorname{Re}(\tilde{s}_i^{\varepsilon,t}) < \operatorname{Re}(\tilde{s}_j^{\varepsilon,t}) < 0, \quad \operatorname{Im}(\tilde{s}_i^{\varepsilon,t}) > \operatorname{Im}(\tilde{s}_j^{\varepsilon,t}) > 0.$$

Therefore one can easily verify that the above condition can be satisfied. Moreover, condition (2.2.2) is satisfied, and this is the reason to impose condition (3.2.1).

Hence the critical values behave for $\varepsilon_0 \geq \varepsilon \geq 0$, $\varepsilon \longrightarrow 0$, as illustrated in Figure 3.2.1. If two flow lines cross each other at some point, then we have indicated in this figure the order in which the critical values pass through this point by interrupting that line which arrives at this point later. The situation at these points is just guaranteed by the above condition.

We now replace the family of functions $g_{\varepsilon,t}: X_t' \longrightarrow \mathbb{C}$, $\varepsilon \in [0,\varepsilon_0]$, by a nearby continous family $\bar{g}_{\varepsilon,t}: X_t' \longrightarrow \mathbb{C}$, $\varepsilon \in [0,\varepsilon_0]$,

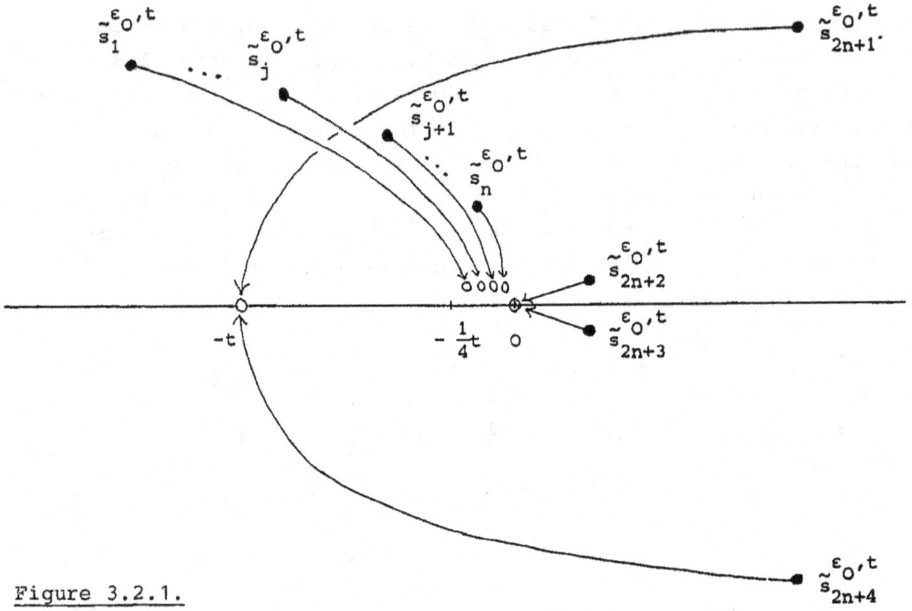

Figure 3.2.1.

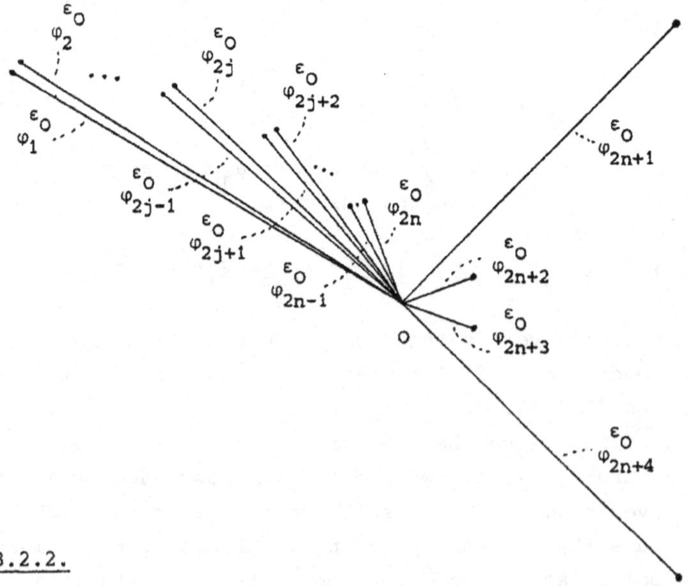

Figure 3.2.2.

of Morse functions with distinct critical values such that for all $\varepsilon \in [0,\varepsilon_0]$ the function $\bar{g}_{\varepsilon,t}$ is close to the function $g_{\varepsilon,t}$. Then to each critical value $\tilde{s}_j^{\varepsilon,t}$ for $j = 1,\dots,n$ of $g_{\varepsilon,t}$ there correspond two critical values $s_{2j-1}^{\varepsilon,t}$ and $s_{2j}^{\varepsilon,t}$ of $\bar{g}_{\varepsilon,t}$ which are close to $\tilde{s}_j^{\varepsilon,t}$, and to each critical value $\tilde{s}_i^{\varepsilon,t}$ for $i = 2n+1,\dots,2n+4$ corresponds one nearby critical value $s_i^{\varepsilon,t}$. We now choose a strongly distinguished system of paths $(\varphi_1^{\varepsilon_0},\dots,\varphi_{2n+4}^{\varepsilon_0})$ from the critical values $s_i^{\varepsilon_0,t}$ for $i = 1,\dots,2n+4$ to the base point $s^{\varepsilon_0} = 0$ as indicated in Figure 3.2.2. This system of paths satisfies the condition (V) of Chapter 2.2.

We now choose a homotopy $(\varphi_1^{\varepsilon},\dots,\varphi_{2n+4}^{\varepsilon})$, $\varepsilon \in [0,\varepsilon_0]$, of the system of paths such that the following conditions are satisfied for $\varepsilon \in [0,\varepsilon_0]$:

(i) The paths φ_i^{ε} connect the critical values $s_i^{\varepsilon,t}$ with the base point s^{ε}.

(ii) The paths φ_i^{ε} are non-self-intersecting.

(iii) Any two paths φ_i^{ε} and φ_j^{ε} for $i \neq j$ have only s^{ε} as a common point.

We also require that the base point $s^{\varepsilon_0} = 0$ passes into the point $s^0 = (\eta,0)$ during the homotopy.

The resulting system of paths $(\varphi_1^0,\dots,\varphi_{2n+4}^0)$ is indicated in Figure 3.2.3.

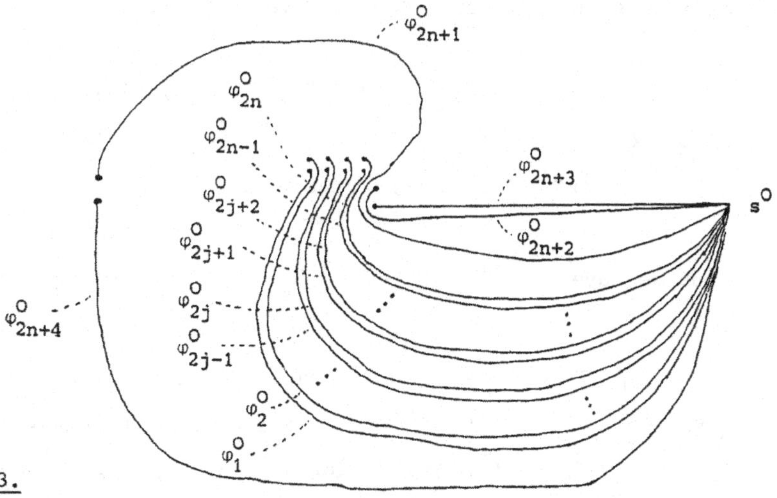

Figure 3.2.3.

This system of paths is related to the system of paths of Chapter 2.3 as follows. The function $g_{0,t}$ has exactly $n+2$ different critical values, each with two ordinary double points lying above it. We choose a strongly distinguished system of paths $(\tilde{\varphi}_1,\ldots,\tilde{\varphi}_{n+2})$ from these critical values to s^0. Let $(\varphi_1,\ldots,\varphi_{2n+4})$ be a strongly distinguished system of nearby paths from the critical values of the nearby Morse function $\bar{g}_{0,t}$ to s^0. According to Remark 2.3.14, a corresponding strongly distinguished basis of thimbles $(\hat{\delta}_1,\ldots,\hat{\delta}_{2n+4})$ then has the intersection matrix of Corollary 2.3.13.

Now we first alter the numbering of the paths by a cyclic permutation

$$(\varphi_1,\varphi_2,\ldots,\varphi_{2n+3},\varphi_{2n+4}) \longrightarrow (\varphi_{2n+4},\varphi_1,\varphi_2,\ldots,\varphi_{2n+3}).$$

Then we apply the transformations

$$\beta_2,\beta_3,\ldots,\beta_{2n+1}$$

to this system of paths. The resulting system of paths is then homotopic to the system of paths $(\varphi_1^0,\ldots,\varphi_{2n+4}^0)$. Therefore one obtains a strongly distinguished basis $(e_1,\ldots,e_{2n+4})$ of thimbles corresponding to this system of paths from $(\hat{\delta}_1,\ldots,\hat{\delta}_{2n+4})$ by changing the numbering correspondingly and applying the above transformations. This yields the following intersection matrix. For n even we have

$$\langle e_i,e_j\rangle = \begin{cases} 0 \text{ for } j=i+1,\ j \text{ even}, & 1\le i,j\le 2n;\ \text{or} \\ & (i,j) = (2n+1,2n+4),\ (2n+2,2n+3) \\ (-1)^{n/2} \text{ for } i<j & \text{otherwise.} \end{cases}$$

For n odd we have

$$\langle e_i,e_j\rangle = \begin{cases} 0 \text{ for } j=i+1,\ j \text{ even},\ 1\le i,j\le 2n; \text{or} (i,j) = (2n+2,2n+3); \\ (-1)^{(n+1)/2}(-1)^{i+j} \text{ for } 1\le i<j\le 2n \text{ otherwise;} \\ (-1)^{(n+1)/2}(-1)^{i+j+1} \text{ for } 1\le i\le 2n,\ 2n+1\le j\le 2n+4; \end{cases}$$

$$\langle e_{2n+1}, e_j \rangle = (-1)^{(n+1)/2}(-1)^{j+1}, \quad j = 2n+2,\ 2n+3\ ;$$

$$\langle e_i, e_{2n+4} \rangle = (-1)^{(n+1)/2}(-1)^{i}, \quad i = 2n+2,\ 2n+3\ ;$$

$$\langle e_{2n+1}, e_{2n+4} \rangle = (-1)^{(n+1)/2}(-2)\ .$$

Applying Theorem 2.2.3 yields a strongly distinguished basis $(e_j^r \mid 1 \le j \le 2n+4,\ 1 \le r \le M_j)$ of thimbles for the singularity $T^n_{2,q,2,s}$ with $M_j = 1$ for $j \neq 2n+2,\ 2n+3$, and $M_{2n+2} = q-1$, $M_{2n+3} = s-1$, and with the following intersection matrix

$$\langle e_j^1, e_{j'}^1 \rangle = \langle e_j, e_{j'} \rangle \text{ for } 1 \le j, j' \le 2n+4\ ;$$

and for $j' = 2n+2,\ 2n+3$:

$$\langle e_j^1, e_{j'}^2 \rangle = \begin{cases} -(-1)^{n(n-1)/2} & \text{for } j = j'\ , \\ -\langle e_{2n+4}^1, e_{j'}^1 \rangle & \text{for } j = 2n+4\ , \\ 0 & \text{otherwise;} \end{cases}$$

$$\langle e_j^r, e_{j'}^{r+1} \rangle = \begin{cases} -(-1)^{n(n-1)/2} & \text{for } j = j'\ ,\ 1 \le r < M_j\ , \\ 0 & \text{otherwise}\ . \end{cases}$$

Now we transform the basis (e_j^r) by the transformations inverse to the above transformations

$$\alpha_{2n}, \alpha_{2n-1}, \dots, \alpha_1\ .$$

These transformations have no effect on the subgraph of the Dynkin diagram corresponding to $\{e_j^r \mid (r,j) > (1, 2n+1)\}$. We denote the new basis by $(\tilde{e}_j^r)$. Then the intersection matrix of $(\tilde{e}_1^1, \dots, \tilde{e}_{2n+4}^1)$ coincides with the intersection matrix of $(\hat{\delta}'_1, \dots, \hat{\delta}'_{2n+4})$ of Chapter 2.3. Now we perform the transformations (B), (C), and (D) (depending on n) indicated there. We denote the resulting strongly distinguished basis of thimbles by $(\tilde{\lambda}_1, \tilde{\lambda}_2, \dots, \tilde{\lambda}_{2n+q+s})$. Then the intersection matrix of $(\tilde{\lambda}_1, \dots, \tilde{\lambda}_{2n+4})$ coincides with the intersection matrix of $(\hat{\delta}''_1, \dots, \hat{\delta}''_{2n+4})$ of Chapter 2.3. We still apply the following transformations, which only alter the numbering:

$\beta_{2n+5};\beta_{2n+7},\beta_{2n+6};\beta_{2n+9},\beta_{2n+8},\beta_{2n+7};\ldots;\beta_{2n+2q-1},\beta_{2n+2q-2},\ldots,\beta_{2n+q+2}$.

In the case when the dimension n is odd we still make the following changes of orientations (cf. the remarks on Figure 2.3.7): First for

$$n = 1 : \quad \varkappa_2,\varkappa_3,\varkappa_4,\varkappa_5;$$

$$n \geq 3 : \quad \varkappa_2,\varkappa_4;\varkappa_{n+3},\varkappa_{n+4},\ldots,\varkappa_{2n};\varkappa_{2n+1},\varkappa_{2n+4};$$

and then in both cases

$$\varkappa_{2n+2j+3}\ , \quad j = 1,\ldots,\left[\frac{q-1}{2}\right]\ ; \quad \varkappa_{2n+q+2j+1}\ , \quad j = 1,\ldots,\left[\frac{s-1}{2}\right]\ .$$

In this way we finally obtain a strongly distinguished basis $(\lambda_1,\lambda_2,\ldots,\lambda_{2n+q+s})$ of thimbles with the Dynkin diagram shown in Figure 3.2.4.

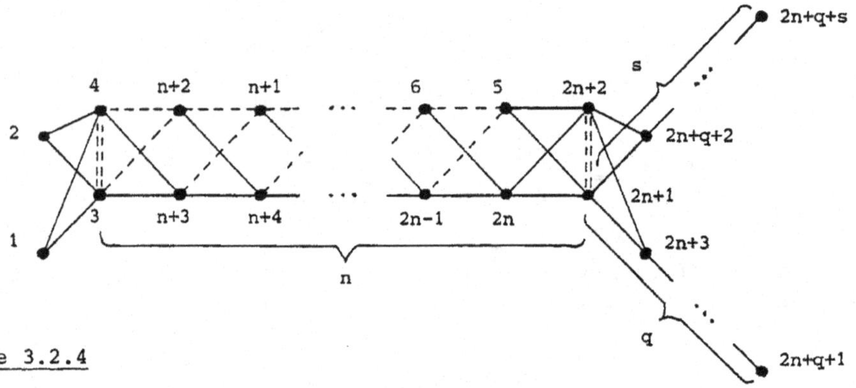

Figure 3.2.4

Let $\Pi^n_{p,q,r,s}$ be the graph of Figure 3.2.5 with $m = 2n + p + q + r + s - 4$. Then the graph of Figure 3.2.4 is strongly equivalent to the graph $\Pi^n_{2,q,2,s}$. Namely, it differs from this graph only by the numbering of the vertices. But this numbering can be altered correspondingly by braid group transformations and changes of orientations. So for example the order of the vertices of the subgraph corresponding to $\{\lambda_{2n+1},\ldots,\lambda_{2n+q+1}\}$ can be totally reversed by the following transformations:

$$\beta_{2n+3},\beta_{2n+2};\beta_{2n+4},\beta_{2n+3};\ldots;\beta_{2n+q+1},\beta_{2n+q};$$

$$\varkappa_{2n+1},\varkappa_{2n+2},\ldots,\varkappa_{2n+q+1}.$$

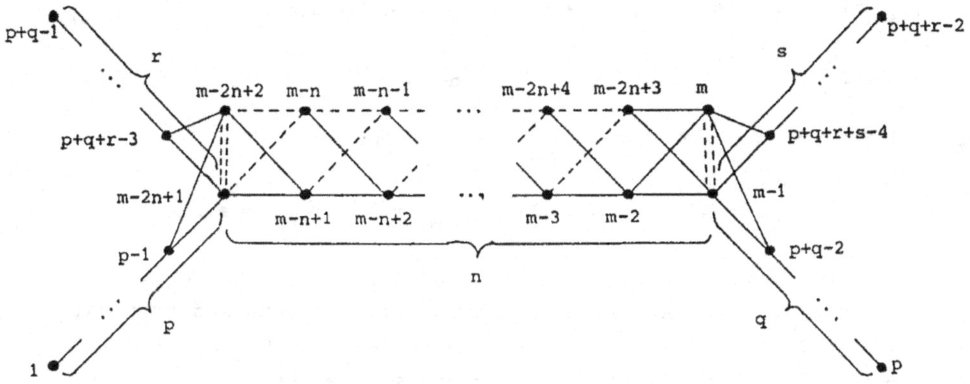

Figure 3.2.5: The graph $\Pi^n_{p,q,r,s}$

(B) Segre symbol $\{1,\ldots,1,2\}$

In this case we use the mapping $f = (f_1,f_2)$ with

$$f_1(z) = z_1^2 + z_2^2 + \ldots + z_n^2 + 2z_{n+1}z_{n+2} ,$$
$$f_2(z) = a_1z_1^2 + a_2z_2^2 + \ldots + a_nz_n^2 + z_{n+1}^2 + z_{n+2}^s$$

for the calculation. Here $3 \le s$, and as in case (A)

$$a_i \in \mathbb{C} , \quad -\frac{1}{4} < \operatorname{Re} a_i < \operatorname{Re} a_j < 0 \text{ for } i < j ,$$
$$0 < \operatorname{Im} a_i = \operatorname{Im} a_j \ll 1 \text{ for } i \le i,j \le n ,$$

where the condition on the imaginary part of a_i will again be fixed later. We again choose the last coordinate function z_{n+2} as the linear function ζ. Then one can easily see that

$$\Sigma_{z_{n+2}}(f_2) = \bigcup_{i=1}^{2n+2} \Sigma_i ,$$

where

$$\Sigma_{2j-\varkappa} = \{z_i = 0 \text{ for } 1 \le i \ne j \le n\,,\ z_j = (-1)^{\varkappa}\sqrt{-2a_j}z_{n+2}\,,\ z_{n+1} = a_j z_{n+2}\}$$

$$\Sigma_{2n+1+\varkappa} = \{z_i = 0 \text{ for } i = 1,\ldots,n\,,\ z_{n+2-\varkappa} = 0\}\,,$$

for $1 \le j \le n$ and $\varkappa \in \{0,1\}$. One easily computes that

$$\rho_i = 2 \text{ for } i = 1,\ldots,2n\,,\ \rho_{2n+1} = 2\,,\ \rho_{2n+2} = s\,.$$

Again $g = (f_1, f_2 + \zeta^2)$ defines a singularity of type $\tilde{D}_{n+3}$. By analogy to the case (A), we study the critical points and critical values of the function $g_{\varepsilon,t}: X'_t \longrightarrow \mathbb{C}$ which tend to 0 for $\varepsilon \longrightarrow 0$ and $t \longrightarrow 0$. The critical points lying on the branches $\Sigma_{2j-\kappa}$ (for $\kappa \in \{0,1\}$ and $j = 1,\ldots,n$) for $t = 0$ approximately have the following critical values (for $\kappa = 0,1$)

$$\tilde{s}_j^{\varepsilon,t} = a_j t - \frac{a_j^2}{1 - a_j^2}\varepsilon^2 + o(\varepsilon^s)\,.$$

The other four critical points are the solutions of the following equations, tending to 0 for $\varepsilon \longrightarrow 0$:

$$z_j = 0 \text{ for } j = 1,\ldots,n\,,$$

$$2z_{n+1}z_{n+2} = t\,,$$

$$4z_{n+2}^4 - 4\varepsilon z_{n+2}^3 - t^2 - 2sz_{n+2}^{s+2} = 0\,.$$

The corresponding critical values are the images of these solutions under the mapping

$$z_{n+2} \longmapsto 2(z_{n+2} - \tfrac{3}{4}\varepsilon)^2 - \tfrac{1}{8}\varepsilon^2 + o(\varepsilon^2)\,.$$

The behaviour of the critical values is illustrated in Figure 3.2.6.

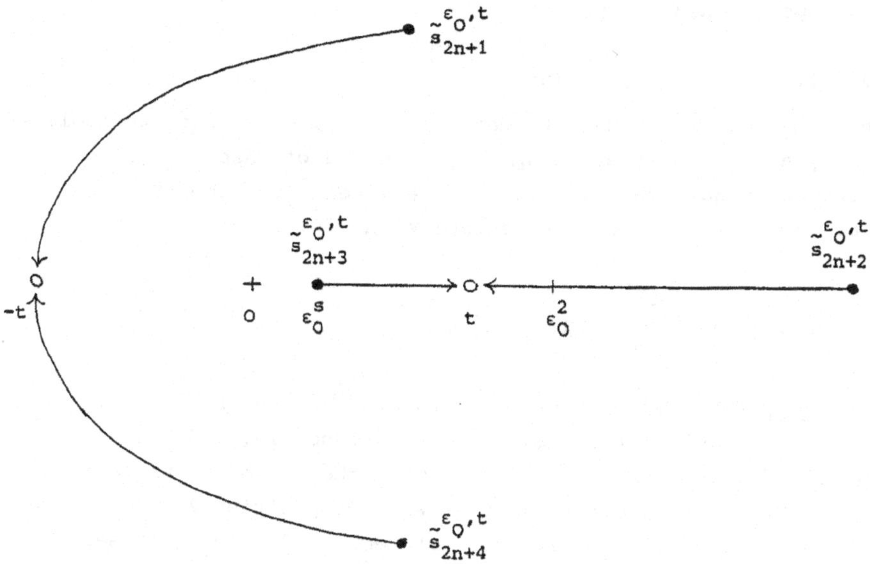

Figure 3.2.6

If we impose the same condition on the imaginary parts of the complex numbers a_i as in case (A), then Figure 3.2.6 has to be completed as in Figure 3.2.1. Proceeding exactly as in case (A), we finally obtain a strongly distinguished basis of thimbles with the Dynkin diagram $\Pi^n_{2,2,2,s}$ of Figure 3.2.5 $(m = 2n + 2 + s)$.

One should compare these results to those of Chapter 2.5. There we obtained the Dynkin diagram of Figure 2.5.4 for the singularity T^n_{2233}. This graph is strongly equivalent to the graph Π^n_{2233}.

3.3. The $J^{(n-1)}$- and $K^{(n-1)}$- series

We now compute Dynkin diagrams of the singularities of the two series of singularities of arbitrary dimension, which were introduced in Section 3.1.

Each of these singularities is given by a mapping of the form

$$f = (w_1^2, a_1 w_1^2) \oplus \ldots \oplus (w_{n-1}^2, a_{n-1} w_{n-1}^2) \oplus \tilde{f} ,$$

where the complex numbers a_i, for $i = 1,\ldots,n-1$, are pairwise distinct and distinct from the eigenvalues of the 2-jet of $\tilde{f}$. By

[Wall_2, Proposition 3.1] one gets

Lemma 3.3.1: $\mu(f) = \mu(\tilde{f}) + 2(n-1)$.

The equation $\tilde{f}(x,y,z) = 0$ defines a space curve with an isolated singularity at 0. In order to apply the method of Chapter 2.2, we choose the coordinate function z as the linear function ζ. Then the polar curve $\Sigma_z(f_2)$ can be decomposed as follows:

$$\Sigma_z(f_2) = \Big(\bigcup_{i=1}^{2n-2} \Sigma_i\Big) \cup \Sigma_z(\tilde{f}_2|_{X' \cap W}),$$

where $X' = f_1^{-1}(0)$ and $W = \{(w;x,y,z) \in \mathbb{C}^{n+2} \mid w_i = 0 \text{ for } i = 1,\dots,n-1\}$. The curves Σ_i, for $i = 1,\dots,2n-2$, are irreducible, and the corresponding invariants ρ_i and ν_i have the following values: $\rho_i = 2$, $\nu_i = 1$. (These results together with Proposition 2.1.4 and Proposition 2.1.5 yield another proof of Lemma 3.3.1 in the cases considered by us.)

As in [Wall_4] we eliminate the variable x from the equations $\tilde{f}_1 = 0$ and $\tilde{f}_2 = 0$. In this way one obtains a function $L_x\tilde{f}$ in the variables y,z, which belongs to one of the series in Arnol'd's classification [Arnol'd_2]. Geometrically, this elimination corresponds to the following operation: The space curve given by $\tilde{f} = 0$ consists of the x-axis and a residual space curve $\tilde{X}$ for the $J^{(n-1)}$-series, respectively only of a space curve $\tilde{X}$ for the $K^{(n-1)}$-series. Then $(L_x\tilde{f})(y,z) = 0$ is just the equation of the projection of $\tilde{X}$ on the (y,z)-plane. The polar curve $\Sigma_z(\tilde{f}_2|_{X' \cap W})$ is then the union of the x-axis with a curve $\tilde{\Sigma}$ whose projection on the (y,z)-plane is just the polar curve $\Sigma_z(L_x\tilde{f})$ of $L_x\tilde{f}$. Let

$$\Sigma_z(L_x\tilde{f}) = \bigcup_{i=1}^{\ell} \Sigma_i^L$$

be the decomposition of $\Sigma_z(L_x\tilde{f})$ into irreducible components Σ_i^L, and let ρ_i^L be the corresponding invariants of Chapter 2.1. Let Σ_{2n-2+i}, $i = 1,\dots,\ell$ be that component of $\tilde{\Sigma}$ which is projected onto Σ_i^L, and let ρ_{2n-2+i} and ν_{2n-2+i} be the corresponding invariants. We shall obtain relations between the numbers ρ_i^L and ρ_{2n-2+i} for both series. Using Proposition 2.1.4 and Proposition 2.1.5, we can then derive a formula which relates the Milnor number $\mu(f)$ to the Milnor number $\mu(L_x\tilde{f})$. Eventually, there will be a relation between the Dynkin diagrams of f and of $L_x\tilde{f}$.

For the computation of Dynkin diagrams we shall again consider the mapping $g = (f_1, f_2 + z^2)$. It defines a singularity of tpye $\tilde{D}_{n+3}$. We shall again study the behaviour of the critical values of $g_{\varepsilon,t} = f_2 + (z-\varepsilon)^2|_{X_t'}$ for $\varepsilon \longrightarrow 0$. Since we are only interested in critical points tending to 0 for $\varepsilon \longrightarrow 0$ and $t \longrightarrow 0$, we can write the coordinates of the critical points as power series in the variables ε and t, and compute the terms of lowest order, which determine the asymptotic behaviour of the critical points.

We shall now enter into details of the calculations.

(A) The $J^{(n-1)}$-series (Segre symbol $\{1,\ldots,1,3\}$)

In this case we assume that the numbers a_i satisfy the following conditions: $a_i \in \mathbb{R}$, $0 < a_i < a_j$ for $i < j$, $1 \leq i,j \leq n-1$.

The curves Σ_{2j-1} and Σ_{2j}, for $j = 1,\ldots,n-1$, are the branches of the curve given by the following equations:

$$x = -a_j^2 z - \frac{1}{2}(\partial Q/\partial y)(y,z) ,$$

$$y = a_j z ,$$

$$w_j^2 = 3a_j^2 z^2 + z(\partial Q/\partial y)(y,z) , \quad w_i = 0 \text{ for } 1 \leq i \neq j \leq n-1 .$$

After eliminating x from the equations $\tilde{f}_1(x,y,z) = 2xz - y^2 = 0$ and $\tilde{f}_2(x,y,z) = 2xy + Q(y,z) = 0$, one gets the function

$$(L_x\tilde{f})(y,z) = 2y^3 + 2zQ(y,z).$$

This is a prenormal form for functions with 3-jet y^3. According to Arnol'd [Arnol'd$_2$], these form the E/J-series. We follow C.T.C. Wall by writing $E_{p,r}$ for Arnol'd's $J_{p,r}$. Then the function-germs with 3-jet y^3 simply form the E-series. The reduction L_x yields a bijection between the μ-constant strata of $\tilde{f}$ and those of the E-series.

The polar curve $\Sigma_z(\tilde{f}_2|_{X'\cap W})$ is given by the equations

$$2xz + 2y^2 + z(\partial Q/\partial y)(y,z) = 0 ,$$

$$2xz - y^2 = 0$$

in W. For the polar curve of $L_x\tilde{f}$ we get

$$\Sigma_z(L_x\tilde{f}) = \{(y,z) \mid 3y^2 + z(\partial Q/\partial y)(y,z) = 0\} .$$

The number ℓ of irreducible components of this curve is equal to 1 or 2, depending on Q. We have

$$\rho_{2n-2+i} = \rho_i^L - 1 \text{ for } i = 1,\dots,\ell ;$$

$$\mu(\tilde{f}_1,\tilde{f}_2-\varepsilon z) = 3 ;$$

$$\nu_1 = 2 \text{ for } \ell = 1 \text{ and } \nu_1 = \nu_2 = 1 \text{ for } \ell = 2 .$$

Thus we obtain the following lemma from Proposition 2.1.4 and Proposition 2.1.5 together with Lemma 3.3.1 (cf. also [Wall$_3$, Lemma 5.1]):

<u>Lemma 3.3.2:</u> $\mu(f) = 2(n-1) + \mu(L_x\tilde{f}) + 1$.

On both branches Σ_{2j-1} and Σ_{2j}, for $j = 1,\dots,n-1$, lies one critical point of $g_{\varepsilon,0}$ different from zero. Asymptotically for $\varepsilon \longrightarrow 0$ and $t \longrightarrow 0$, the corresponding critical points of $g_{\varepsilon,t}$ have the critical value

$$a_j t + \frac{a_j^3}{a_j^3+1}\varepsilon^2 .$$

For the critical points of $g_{\varepsilon,t}$ lying on $\Sigma_z(\tilde{f}_2|_{X'\cap W})$ we have: After eliminating x from the equations for the critical locus of $g_{\varepsilon,t}$, one obtains the following equations for the terms of lowest order of the coordinates of these points, written as power series in ε and t:

$$y^2 = -\frac{1}{3}t ,$$

$$(z^2(z-\varepsilon) + \sqrt{-1}(\sqrt{t/3})^3)(z^2(z-\varepsilon) - \sqrt{-1}(\sqrt{t/3})^3) = 0 \qquad (*)$$

The terms of lowest order of the corresponding critical values are the images of the solutions of these equations under the mapping

$$z \longmapsto 3(z - \tfrac{2}{3}\varepsilon)^2 - \tfrac{1}{3}\varepsilon^2 ,$$

except for $z = \varepsilon$. In this case the asymptotics of the critical values are given by Proposition 2.2.1 and depend on Q. According to the conditions on the orders of the terms of Q, they have order $o(\varepsilon^2)$. Thus the above approximation is also valid for $z = \varepsilon$ up to order $o(\varepsilon^2)$.

Now one computes that the behaviour of the critical values of $g_{\varepsilon,t}$ for $t \neq 0$ and $\varepsilon = \varepsilon_0 \longrightarrow \varepsilon = 0$ can be illustrated as in Figure 3.3.1. This figure is based on a computer graphic which was plotted by a home computer (Schneider CPC 6128) by means of a simple program in BASIC. This program makes use of the fact that for a fixed t and a fixed argument of z there are at most two solutions (z,ε) of the equation (*) with $0 \leq \varepsilon \leq \varepsilon_0$, which can be written down explicitly.

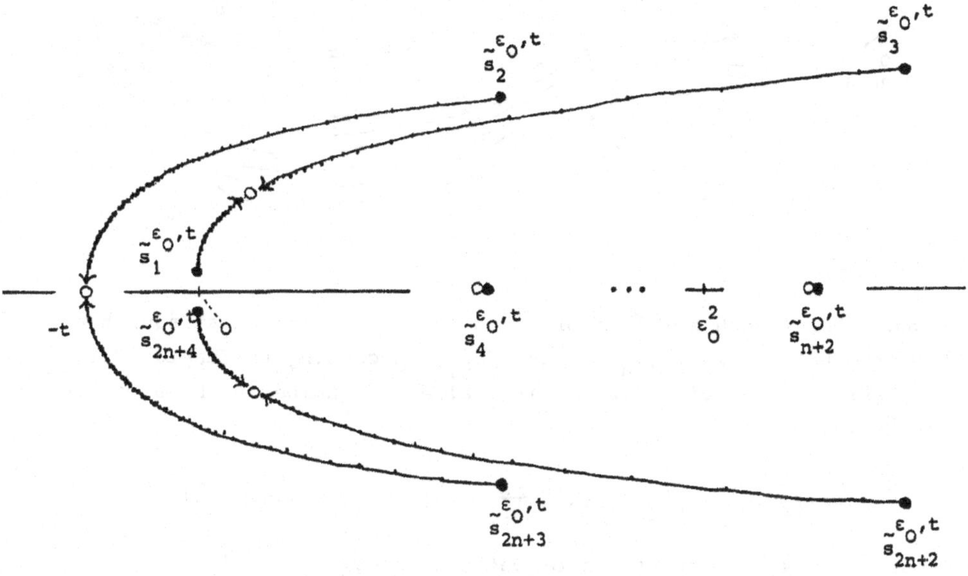

Figure 3.3.1

As in Section 3.2, we now replace the family of functions $g_{\varepsilon,t}$ for $\varepsilon \in [0,\varepsilon_0]$ by a continuous family of nearby Morse functions $\bar{g}_{\varepsilon,t}$ with different critical values. Then the critical values $\tilde{s}_j^{\varepsilon,t}$ split into two critical values $s_{2j-4}^{\varepsilon,t}$ and $s_{2j-3}^{\varepsilon,t}$ each, for $j = 4,\ldots,n+2$. We choose a strongly distinguished system of paths $(\varphi_1^{\varepsilon_0},\ldots,\varphi_{2n+4}^{\varepsilon_0})$ from the critical values of $g_{\varepsilon_0,t}$ to the base point $s^{\varepsilon_0} = 0$ in the given order. The paths $\varphi_i^{\varepsilon_0}$, $i = 1,2,3,2n+2,2n+3,2n+4$, are simply chosen as line segments. The remaining paths have to be smooth, and the real parts of the tangent vectors in each point must be negative. Then condition (V) is satisfied. As in Section 3.2, we choose a homotopy $(\varphi_1^\varepsilon,\ldots,\varphi_{2n+4}^\varepsilon)$, $\varepsilon \in [0,\varepsilon_0]$ of the system of paths with the properties indicated there. The resulting system of paths $(\varphi_1^0,\ldots,\varphi_{2n+4}^0)$ is illustrated in Figure 3.3.2.

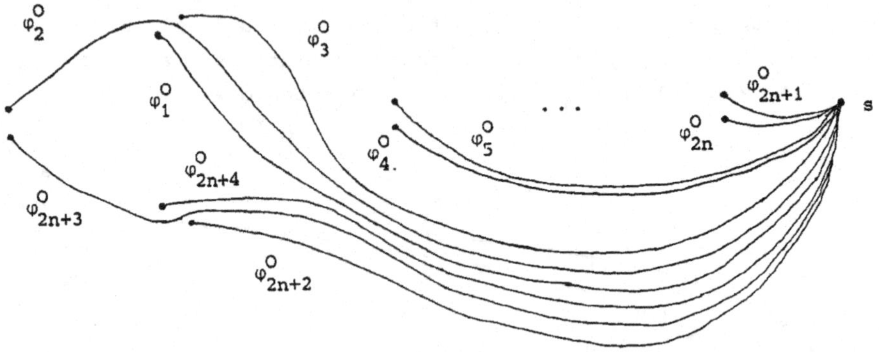

Figure 3.3.2

One obtains such a system of paths from a strongly distinguished system of paths $(\varphi_1,\ldots,\varphi_{2n+4})$ for $\bar{g}_{0,t}$ according to Chapter 2.3 as follows: As in Section 3.2 we first alter the numbering by a cyclic permutation

$$(\varphi_1,\varphi_2,\ldots,\varphi_{2n+3},\varphi_{2n+4}) \longrightarrow (\varphi_{2n+4},\varphi_1,\varphi_2,\ldots,\varphi_{2n+3}) \quad .$$

By Remark 2.3.14 the new system of paths defines a strongly distinguished basis of thimbles with the same intersection matrix (after an appropriate orientation of the thimbles) as the intersection matrix of the basis $(\hat{\delta}_1',\ldots,\hat{\delta}_{2n+4}')$ of Chapter 2.3. The above system of paths $(\varphi_1^0,\ldots,\varphi_{2n+4}^0)$ can be obtained from such a system of paths (up to homotopy) by the transformations

$$\beta_2,\alpha_{2n+3} \quad .$$

Let $(\tilde{e}_1,\ldots,\tilde{e}_{2n+4})$ be the strongly distinguished basis of thimbles for g resulting from the above basis by applying the transformations β_2 and α_{2n+3}. By Theorem 2.2.3 this basis determines a strongly distinguished basis $(\tilde{e}_j^r)$ of thimbles of f. We apply the transformations α_1 and β_{2n+4}, which are inverse to the above transformations, to this basis. Then one finally obtains a strongly distinguished basis of thimbles $(e_j^r \mid 1 \le j \le 2n+4,\ 1 \le r \le M_j)$ for f with the following properties:

(i) The intersection matrix of $(e_1^1, \ldots, e_{2n+4}^1)$ coincides with the intersection matrix of $(\hat{\delta}_1', \ldots, \hat{\delta}_{2n+4}')$.

(ii) The complete intersection matrix is related to the intersection matrix of $(e_1^1, \ldots, e_{2n+4}^1)$ by the formulas of Theorem 2.2.3.

The numbers M_j have the following values:

$$M_j = 1 \quad \text{for} \quad j \neq 2, 2n+3, \quad M_2 = M_1^L - 1, \quad M_{2n+3} = M_2^L - 1,$$

where M_1^L, M_2^L are the corresponding numbers of Table 1 of [Gabrielov$_3$] for the singularity $L_x\tilde{f}$. In detail one gets the following table, where we have listed the equation of $\tilde{f}$ used for the computation in each case.

Table 3.3.1.

Notation	Mapping $\tilde{f}$	M_2, M_{2n+3}
$J_{6k+2n+5}^{(n-1)}$	$(2xz-y^2, 2xy+z^{3k+3})$	3k+2,3k+2
$J_{6k+2n+6}^{(n-1)}$	$(2xz-y^2, 2xy+yz^{2k+2})$	3k+3,3k+2
$J_{6k+2n+7}^{(n-1)}$	$(2xz-y^2, 2xy+z^{3k+4})$	3k+3,3k+3
$J_{k+1,i}^{(n-1)}$	$(2xz-y^2, 2xy+y^2z^k+z^{3k+2+i})$	3k+i+1,3k+1

(B) The $K^{(n-1)}$-series (Segre symbol $\{1, \ldots, 1, (1,2)\}$)

In this case we make the same assumptions for the numbers a_i as in Section 3.2.

The curves Σ_{2j-1} and Σ_{2j} for $j = 1, \ldots, n-1$ are the branches of the curve given by the equations

$$x = a_j z,$$

$$a_j y = \frac{1}{2}(\partial Q/\partial y)(y,z),$$

$$w_j^2 = -2a_j z^2 - y^2, \quad w_i = 0 \quad \text{for} \quad 1 \leq i \neq j \leq n-1.$$

After eliminating x from the equations $\tilde{f}_1(x,y,z) = 2xz + y^2 = 0$, $\tilde{f}_2(x,y,z) = x^2 + Q(y,z) = 0$, one gets the function

$$(L_x \tilde{f})(y,z) = y^4 + 4z^2 Q(y,z).$$

This is a prenormal form for functions with 4-jet y^4. Hence the function $L_x \tilde{f}$ belongs to one of the series W, X, Y, or Z by [Arnol'd$_2$], and we get a bijection between the μ-constant strata of $\tilde{f}$ and those of these series.

The polar curve $\Sigma_z(\tilde{f}_{2|X' \cap W})$ is given by the equations

$$2xy - z(\partial Q/\partial y)(y,z) = 0,$$

$$y^2 + 2xz = 0$$

in W, and the polar curve of $L_x \tilde{f}$ by

$$y^3 + z^2(\partial Q/\partial y)(y,z) = 0.$$

The number ℓ of irreducible components of this curve is equal to 1,2, or 3. By considering the Puiseux parametrizations of these components, one can easily show :

$$\rho_{2n-2+i} = \rho_i^L - 2 \quad \text{for} \quad i = 1,\dots,\ell .$$

Moreover, we have

$$\sum_{i=1}^{\ell} \nu_i = 3, \quad \mu(\tilde{f}_1, \tilde{f}_2 - \varepsilon z) = 2.$$

In this way one can deduce from Proposition 2.1.4 and Proposition 2.1.5 together with Lemma 3.3.1 (cf. also [Wall$_3$, Lemma 5.2]):

Lemma 3.3.3: $\mu(f) = 2(n-1) + \mu(L_x \tilde{f}) - 4$

On both branches Σ_{2j-1} and Σ_{2j}, for $j = 1,\dots,n-1$, lies one critical point of $g_{\varepsilon,0}$ different from zero. Asymptotically for $\varepsilon \longrightarrow 0$ and $t \longrightarrow 0$, the corresponding critical points of $g_{\varepsilon,t}$ have the critical value

$$a_j t - \frac{a_j^2}{1 - a_j^2}\varepsilon^2 \quad .$$

For the critical points of $g_{\varepsilon,t}$ lying on $\Sigma_z(\tilde{f}_2|_{X'\cap W})$ for $t = 0$, one computes the following results. Four of these critical points have critical values which asymptotically coincide with the approximate values for the critical values $\tilde{s}_i^{\varepsilon,t}$ for $i = 2n+1,\dots,2n+4$ of Section 3.2 (B) (cf. Figure 3.2.6). In addition one has two further critical points approximately given by $(x,y,z) = (\pm\sqrt{t},0,\varepsilon)$ with corresponding critical values having at least order $o(\varepsilon^2)$, where the exact values depend on Q. Up to these two additional critical values and the missing critical value $\tilde{s}_n^{\varepsilon,t}$, one therefore has exactly the situation of Section 3.2 (B).

An analogous calculation as in Section 3.2 (B) then yields a strongly distinguished basis $(e_j^r \mid 1 \le j \le 2n+4,\ 1 \le r \le M_j)$ of thimbles for f which has the same properties as the corresponding basis of the $J^{(n-1)}$-series, with

$$M_j = 1 \quad \text{for} \quad j \neq 2, 2n+2, 2n+3,$$

$$M_2 = M_2^L - 2,\quad M_{2n+2} = M_3^L - 2,\quad M_{2n+3} = M_1^L - 2,$$

where M_1^L, M_2^L, and M_3^L are the corresponding numbers for the singularity $L_x\tilde{f}$. (In fact the calculation yields $M_2 = 1$ and $M_{2n+1} = M_2^L - 2$, but by the transformation α_{2n}, followed by transformations as in Remark 2.3.14 and possibly changes of orientations, one can obtain the above basis whose numbering corresponds to the numbering in the case of the $J^{(n-1)}$-series.)

In particular one can deduce Table 3.3.2 from Table 1 of [Gabrielov$_3$]. Here we restrict ourselves to those mappings f with reduction $L_x\tilde{f}$ belonging to the W-series. (By [Wall$_3$] one can find K-unimodal map-germs only among these, and indeed only for $n = 1,2$.) We adopt the notation of [Wall$_3$].

Table 3.3.2.

Notation	Mapping $\tilde{f}$	M_2, M_{2n+2}, M_{2n+3}
$\cdot K^{(n-1)}_{12k+2n-6}$	$(2xz+y^2, x^2+z^{4k-1})$	4k-2,4k-2,4k-2
$K^{(n-1)}_{12k+2n-5}$	$(2xz+y^2, x^2-yz^{3k-1})$	4k-2,4k-2,4k-1
$K^{(n-1)}_{k,i}$	$(2xz+y^2, x^2-y^2z^{2k-1}+z^{4k+i})$	4k+i-1,4k-1,4k-1
$K^{(n-1),\#}_{k,2q-1}$	$(2xz+y^2, x^2-\frac{1}{2}y^2z^{2k-1}+yz^{3k+q-1}+\frac{1}{4}z^{4k})$	4k-1,4k+q-1,4k+q-2
$K^{(n-1),\#}_{k,2q}$	$(2xz+y^2, x^2-\frac{1}{2}y^2z^{2k-1}+y^2z^{2k+q-1}+\frac{1}{4}z^{4k})$	4k-1,4k+q-1,4k+q-1
$K^{(n-1)}_{12k+2n-1}$	$(2xz+y^2, x^2-yz^{3k})$	4k,4k,4k-1
$K^{(n-1)}_{12k+2n}$	$(2xz+y^2, x^2+z^{4k+1})$	4k,4k,4k

We now consider those singularities of Tables 3.3.1 and 3.3.2 which are at the bottom of the hierarchy of the corresponding series. This means that all the other singularities of one of these series deform into one of these singularities. These are the singularities $J^{(n-1)}_{2n+5}$, $J^{(n-1)}_{2n+6}$, $J^{(n-1)}_{2n+7}$ (k = 0), $K^{(n-1)}_{2n+6}$, and $K^{(n-1)}_{2n+7}$ (k = 1). The Dynkin diagrams of these singularities can be transformed to a "normal form" extending the graph $\Pi^n_{2,q,2,s}$.

<u>Proposition 3.3.4.</u> <u>The singularity</u> $J^{(n-1)}_{2n+2+s}$, s = 3,4,5, <u>has a strongly distinguished basis of thimbles with the Dynkin diagram</u> $\tilde{\Pi}^n_{2,2,2,s}$, <u>and the singularity</u> $K^{(n-1)}_{2n+3+s}$, s = 3,4, <u>has a corresponding basis with Dynkin diagram</u> $\tilde{\Pi}^n_{2,3,2,s}$. <u>Here</u> $\tilde{\Pi}^n_{p,q,r,s}$ <u>is the graph shown in Figure 3.3.3 with</u> $m = 2n+p+q+r+s-3$.

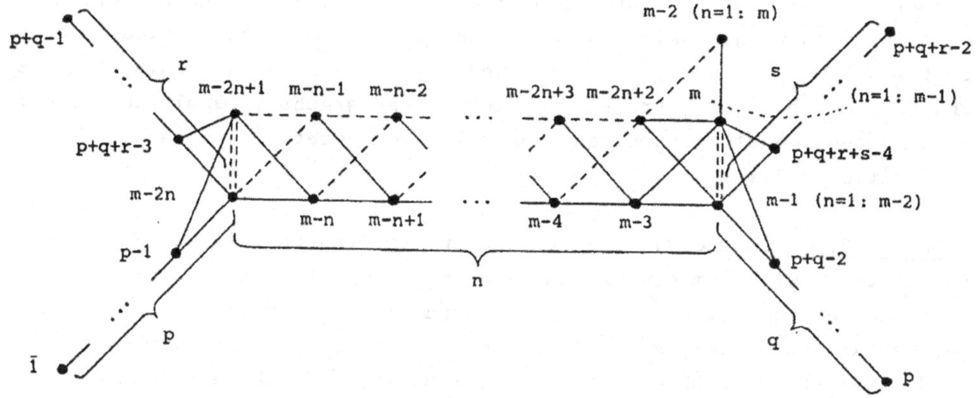

<u>Figure 3.3.3:</u> The graph $\tilde{\Pi}_{p,q,r,s}$

<u>Proof.</u> We only consider the cases $J_{2n+5}^{(n-1)}$ and $K_{2n+6}^{(n-1)}$. The other cases can be reduced to these cases similarly as in [Ebeling$_1$]. For both of these singularities we above obtained a basis $(e_j^r \mid 1 \leq j \leq 2n+4,$ $1 \leq r \leq M_j) =: (e_1,\ldots,e_m)$ where the intersection matrix of $(e_1,\ldots,e_{2n+4}) = (e_1^1,\ldots,e_{2n+4}^1)$ coincides with the intersection matrix of $(\hat{\delta}_1',\ldots,\hat{\delta}_{2n+4}')$ of Chapter 2.3 for the singularity $\tilde{D}_{n+3}$. We again apply the transformations (B), (C), and (D) of Chapter 2.3 to this basis, where, however, the subsequence (B) of transformations is slightly altered. We replace the transformations of (B) by the following transformations:

$$\left.\begin{array}{l} \alpha_{2n-2},\alpha_{2n-1},\alpha_{2n}; \\ \alpha_{2n-4},\alpha_{2n-3},\alpha_{2n-2},\alpha_{2n-1},\alpha_{2n}; \\ \vdots \\ \alpha_2,\alpha_3,\ldots,\alpha_{2n}; \end{array}\right\} \text{ for } n \geq 2 \text{ only}$$

$$\alpha_1,\alpha_2,\ldots,\alpha_{2n};$$

$$\alpha_{2n+4}^{(*)},\alpha_{2n+3}^{(*)},\alpha_{2n+2}^{(*)};\alpha_{2n+4},\alpha_{2n+3}.$$

Up to the transformations denoted by (*), these are transformations of the underlying Dynkin diagram of $\tilde{D}_{n+3}$. They coincide with the transformations (B) up to transformations which only alter the numbering.

The resulting Dynkin diagrams then coincide with the indicated graphs of type $\tilde{\Pi}^n_{p,q,r,s}$ up to numbering and orientation of the basis elements, and they can be easily transformed into these graphs by applying further braid group transformations and changes of orientations. This proves Proposition 3.3.4.

<u>Remark 3.3.5.</u> Figure 3.3.3 should be interpreted as follows. For $n = 1$ the vertices with numbers $m - 2n$ and $m - 2n + 1$ have to be identified with the vertices with numbers $m - 2$ and $m - 1$ respectively. In particular the vertex m is only connected with the vertex $m - 1$. For $n \geq 2$, however, a graph of the form of Figure 3.3.3, where the edge between the vertex $m - 2$ and the vertex $m - 2n + 2$ is omitted, is in general not a Dynkin diagram corresponding to a <u>strongly</u> distinguished basis of thimbles. This follows from the fact that e.g. for $n = 2$ and $(p,q,r,s) = (2,2,2,3)$ the characteristic polynomial of the corresponding Coxeter element is not a product of cyclotomic polynomials. However, such a graph is <u>weakly</u> equivalent to the graph $\tilde{\Pi}^n_{p,q,r,s}$. For by performing the transformations

$$\alpha_m(m - 2n + 2), \alpha_{m-1}(m - 2n + 2),$$

a basis with a Dynkin diagram $\tilde{\Pi}^n_{p,q,r,s}$ can be transformed into a weakly distinguished basis $(\lambda'_1,\ldots,\lambda'_m)$ with the same Dynkin diagram except that $\langle\lambda'_{m-2},\lambda'_{m-2n+2}\rangle = 0$. But this means that the basis $(\lambda_1,\ldots,\lambda_m)$ with

$$\lambda_1 = \lambda'_{m-2},\ \lambda_2 = \lambda'_m,\ \lambda_3 = \lambda'_{m-1}$$

and $\{\lambda_1,\ldots,\lambda_m\} = \{\lambda'_1,\ldots,\lambda'_m\}$ is a special basis in the sense of Definition 5.4.1. For the corresponding lattice $\hat{H}$ one gets

$$\hat{H} = Q^n_{pqrs} \perp U \perp \ker \hat{H},\ \dim(\ker \hat{H}) = n - 1,$$

where the lattice Q^n_{pqrs} is defined in Chapter 4.1. In particular, a comparison with the results of Chapter 2.5 shows that the lattices $\hat{H}$ coincide for the singularities T^n_{2223} and $J^{(n-1)}_{2n+5}$.

In the following sections we shall study the cases $n = 1$ and $n = 2$ more precisely.

3.4. Simple space curves

We shall now consider the case of dimension $n = 1$, hence the case of space curve singularities. We want to study the Dynkin diagrams of the simple (0-modal) space curve singularities. An isolated complete intersection singularity (X,x) is called simple (0-modal), if only finitely many isomorphism classes of singularities occur in the semi-universal deformation of (X,x). The simple isolated complete intersection singularities were classified by M. Giusti [Giusti$_1$]. (This classification is implicitly already contained in [Mather$_1$].) In dimension $n \geq 1$ there are besides the simple hypersurface singularities only simple space curve singularities, more precisely only singularities of map-germs $f : (\mathbb{C}^3,0) \longrightarrow (\mathbb{C}^2,0)$. These are the following singularities:

$$S_5 = \tilde{D}_4 ,$$

$$S_\mu = T^1_{2,2,2,\mu-3} , \quad \mu \geq 6$$

$$T_\mu = T^1_{2,3,2,\mu-4}, \quad \mu = 7,8,9;$$

$$U_\mu = J_\mu, \quad \mu = 7,8,9;$$

$$W_\mu = K_\mu, \quad \mu = 8,9 ,$$

as well as the singularities Z_9 and Z_{10}, whose 2-jet is singular. Here we contrasted the notation of Giusti with Wall's, respectively our, notation.

Thus the simple space curve singularities with regular 2-jet are already contained in the classes of singularities considered up to now, and we have already got Dynkin diagrams for them. In particular one can find Dynkin diagrams of the form $\Pi^1_{p,q,r,s}$ (Figure 3.2.5) or $\tilde{\Pi}^1_{p,q,r,s}$ (Figure 3.3.3) for them. We now consider a somewhat more general class of graphs, which includes these graphs. The general graph of this class is shown in Figure 3.4.1.

We denote the graph of Figure 3.4.1 by $\tilde{\Theta}_{p_1,\ldots,p_h}$. The subgraph with the vertex with number ρ omitted is denoted by $\Theta_{p_1,\ldots,p_h}$. One has

$$\rho = \sum_{i=1}^{h} p_i - h + 3.$$

Then in particular $\Pi^1_{p,q,r,s} = \Theta_{p,q,r,s}$ and $\tilde{\Pi}^1_{p,q,r,s} = \tilde{\Theta}_{p,q,r,s}$. According to Section 3.2 and Proposition 3.3.4 one has the following result.

Proposition 3.4.1. For a simple space curve singularity given by a map-germ $f : (\mathbb{C}^3,0) \longrightarrow (\mathbb{C}^2,0)$ with a regular 2-jet, there exists a strongly distinguished basis of thimbles with a Dynkin diagram of the form $\Theta_{p,q,r,s}$ or $\tilde{\Theta}_{p,q,r,s}$ as indicated in Table 3.4.1.

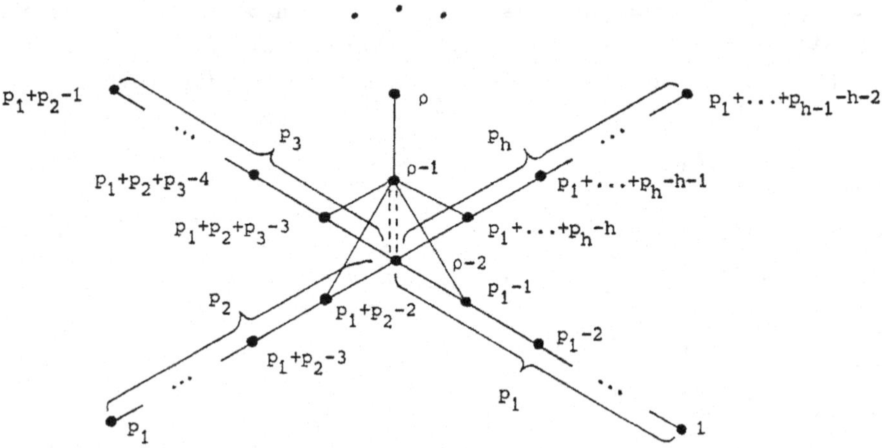

Figure 3.4.1: The graphs $\Theta_{p_1,\ldots,p_h}$ and $\tilde{\Theta}_{p_1,\ldots,p_h}$.

The simple space curve singularities have quasi-homogeneous equations. However, it is not possible in any case to find quasi-homogeneous equations with f_1 being non-degenerate. Nevertheless, the monodromy operator c and the relative monodromy operator $\hat{c}$ both have finite order N in each case, and the number N coincides with the greater degree of the quasi-homogeneous equations (cf. Theorem 1.6.5). We have indicated in Table 3.4.1 for each singularity the notation of Giusti; Wall's, respectively our, notation; the quasi-homogeneous type $(w_1,w_2,w_3;d_1,d_2)$, where the numbers w_i denote the weights and the numbers d_j the degrees of the equations; the Dynkin diagram; the

Coxeter number N; and the type of the vanishing lattice in Janssen's classification [Janssen$_2$].

Table 3.4.1.

Notation		$(w_1,w_2,w_3;d_1,d_2)$	Dynkin diagram	N	Vanishing lattice (H,Δ)
S_5	$\tilde{D}_4$	(1,1,1;2,2)	Θ_{2222}	2	$A^{odd}(1;3;0)$
S_{3+i}, $i \geq 3$	$T^1_{2,2,2,i}$	$\left\{\begin{matrix}(\frac{i}{2},\frac{i}{2},1;\frac{i}{2}+1,i)\\(i,i,2;i+2,2i)\end{matrix}\right\}$	$\Theta_{2,2,2,i}$	$\left\{\begin{matrix}i\\2i\end{matrix}\right.$	$A^{odd}(1,\dots,1;3;\infty)$ i even $A^{ev}(1,\dots,1;2)$ i odd
T_7	T^1_{2323}	(3,2,2;4,6)	Θ_{2233}	6	$O^{\#}_1(1,1,1;1)$
T_8	T^1_{2324}	(6,4,3;7,12)	Θ_{2234}	12	$O^{\#}(1,1,1;2;\infty)$
T_9	T^1_{2325}	(15,10,6;16,30)	Θ_{2235}	30	$O^{\#}_0(1,1,1,1;1)$
U_7	J_7	(5,4,3;8,9)	$\tilde{\Theta}_{2223}$	9	$O^{\#}_1(1,1,1;1)$
U_8	J_8	(4,3,2;6,7)	$\tilde{\Theta}_{2224}$	7	$O^{\#}(1,1,1;2;\infty)$
U_9	J_9	(7,5,3;10,12)	$\tilde{\Theta}_{2225}$	12	$O^{\#}_0(1,1,1,1;1)$
W_8	K_8	(5,6,4;10,12)	$\tilde{\Theta}_{2233}$	12	$O^{\#}_0(1,1,1,1;0)$
W_9	K_9	(4,5,3;8,10)	$\tilde{\Theta}_{2234}$	10	$O^{\#}(1,1,1,1;1,\infty)$

Remark 3.4.2. Giusti [Giusti$_1$] also gives Dynkin diagrams for the simple space curve singularites, which, however, have only μ and not $m = \mu + 1$ vertices. The relation between these graphs is the following. If one applies to a graph $\tilde{\Theta}_{2,2,r,s}$ (or $\Theta_{2,2,r,s}$) the transformations

$$\beta_3,\beta_4,\dots,\beta_{\rho-1};\beta_{\rho-1};\alpha_1,\alpha_2,\dots,\alpha_{\rho-4};\kappa_\rho,\kappa_{\rho-3},$$

then the resulting graph has the form of Figure 3.4.2.

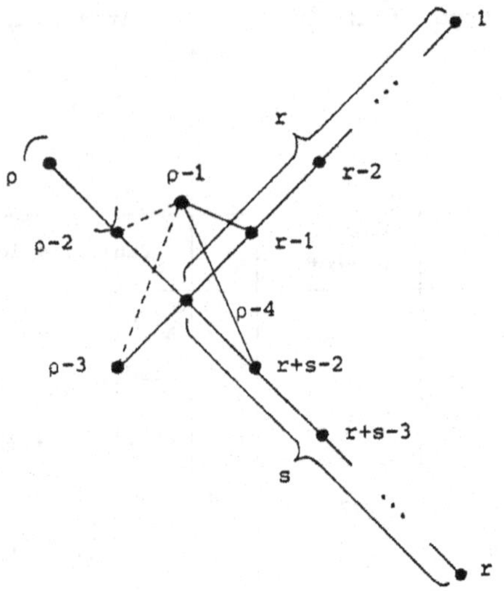

Figure 3.4.2

Let $(\lambda_1',\ldots,\lambda_{\rho-1}'(\lambda_\rho'))$ be the corresponding strongly distinguished basis of $\hat{H}$. Then $\lambda_{\rho-1}' - \lambda_{\rho-4}' \in \ker \hat{H}$. The Dynkin diagram of Giusti is the graph which one obtains after omitting the vertex corresponding to $\lambda_{\rho-1}'$. Giusti's diagram is therefore a Dynkin diagram corresponding to a basis of vanishing cycles of H which can be extended to a strongly distinguished system of generators of vanishing cycles.

We shall now study the general graphs $\Theta_{p_1,\ldots,p_h}$ and $\tilde{\Theta}_{p_1,\ldots,p_h}$ of Figure 3.4.1 in more detail. First we shall make a general remark.

Let D be a Dynkin diagram with vertices $(\lambda_1,\ldots,\lambda_\rho)$. Then one can assign various intersection matrices and Coxeter elements to this graph as follows. Let $\varepsilon = (-1)^{n(n+1)/2}$. Let $V^{(n)} = V^{(n)}(D)$ be the $\rho\times\rho$-matrix $((v_{ij}^{(n)}))$ with

$$v_{ij}^{(n)} = \begin{cases} \varepsilon & \text{for} \quad i = j\,, \\ 0 & \text{for} \quad i > j\,, \\ \pm\varepsilon\,|v_{ij}^{(n)}| & \text{for} \quad i < j\,, \end{cases}$$

where $|v_{ij}^{(n)}|$ for $i<j$ is the number of edges between the vertices λ_i and λ_j and the sign $\pm\varepsilon$ depends on the weight of the edges ($+\varepsilon$ if the edges are dotted, cf. Chapter 1.5). We set

$$A^{(n)} = A^{(n)}(D) = V^{(n)} + (-1)^n (V^{(n)})^t\,,$$
$$\hat{C}^{(n)} = \hat{C}^{(n)}(D) = (-1)^{n+1}(V^{(n)})^{-1}(V^{(n)})^t\,.$$

Then $\hat{C}^{(n)}$ is the matrix of the Coxeter element $\hat{c}^{(n)}$ corresponding to a basis $(\lambda_1^{(n)},\ldots,\lambda_\rho^{(n)})$ of a lattice $\hat{L}^{(n)}$ with intersection matrix $A^{(n)}$ (cf. Proposition 1.6.3). Hence the matrices $\hat{C}^{(n)}$ and $\hat{C}^{(n-1)}$ differ only by a sign, and the matrices $\hat{C}^{(n)}$ and $\hat{C}^{(n-2)}$ are equal. In particular the matrices $\hat{C}^{(n)}$ and $\hat{C}^{(n-1)}$ have the same eigenvalues up to sign, and their Jordan normal forms are equal up to the signs of the eigenvalues. We write

$$P^{(n)}(D)(t) = \det(t.\mathbb{1} - \hat{C}^{(n)})$$

for the characteristic polynomial of $\hat{C}^{(n)}$.

Now let D be one of the two graphs defined by Figure 3.4.1, hence $D = \Theta_{p_1,\ldots,p_h}$ or $D = \tilde{\Theta}_{p_1,\ldots,p_h}$, where $2 \leq p_1 \leq \ldots \leq p_h$. We first consider the lattice $\hat{L}^{(2)}(D)$ defined by the intersection matrix $A^{(2)}(D)$ corresponding to a basis $(\lambda_1^{(2)},\ldots,\lambda_{\rho-1}^{(2)})$, respectively $(\lambda_1^{(2)},\ldots,\lambda_\rho^{(2)})$. Let $Q_{p_1,\ldots,p_h}$ be the lattice spanned by the vectors $\lambda_1^{(2)},\ldots,\lambda_{\rho-2}^{(2)}$. In the case $D = \Theta_{p_1,\ldots,p_h}$ the vector $\lambda_{\rho-1}^{(2)} - \lambda_{\rho-2}^{(2)}$ lies in $\ker \hat{L}^{(2)}(D)$, and one has

$$\hat{L}^{(2)}(\Theta_{p_1,\ldots,p_h}) = Q_{p_1,\ldots,p_h} \perp \mathbb{Z}.(\lambda_{\rho-1}^{(2)} - \lambda_{\rho-2}^{(2)}) .$$

In the case $D = \tilde{\Theta}_{p_1,\ldots,p_h}$ the basis $(\lambda_\rho^{(2)},\lambda_{\rho-1}^{(2)},\ldots,\lambda_1^{(2)})$ forms a special basis in the sense of Definition 5.4.1. Hence

$$\hat{L}^{(2)}(\tilde{\Theta}_{p_1,\ldots,p_h}) = Q_{p_1,\ldots,p_h} \perp U .$$

We now examine the lattice $Q_{p_1,\ldots,p_h}$. By means of the formula in the proof of [Ebeling$_1$, Satz 2.1.1] one can prove the following formula for the corresponding discriminant (i.e. the determinant of the intersection matrix)

$$\operatorname{disc}(Q_{p_1,\ldots,p_h}) = (-1)^\rho \prod_{j=1}^{h} p_j \left(\sum_{i=1}^{h} \frac{1}{p_i} - (h-2) \right) .$$

Let $k_0 = \dim(\ker Q_{p_1,\ldots,p_h})$, and let k_+, respectively k_-, be the dimension of a maximal positive, respectively negative, definite subspace of $Q_{p_1,\ldots,p_h} \otimes \mathbb{R}$. Then we have the following equivalences:

$$(-1)^{\rho}\operatorname{disc}(Q_{p_1,\ldots,p_h}) \begin{matrix} > \\ = \\ < \end{matrix} 0 \iff \sum_{i=1}^{h} \frac{1}{p_i} \begin{matrix} > \\ = \\ < \end{matrix} h-2 \iff (k_0,k_+) = \begin{matrix} (0,0) \\ (1,0) \\ (0,1) \end{matrix} .$$

Hence $k_0 = k_+ = 0$ is true if and only if the Dynkin diagram corresponding to $Q_{p_1,\ldots,p_h}$ is one of the classical Dynkin diagrams $A_{\rho-2}, D_{\rho-2}, E_6, E_7$, or E_8; one has $k_0 = 1$, $k_+ = 0$ if and only if $(p_1,\ldots,p_h) = (2,3,6), (2,4,4), (3,3,3), (2,2,2,2)$; and $k_0 = 0$, $k_+ = 1$ otherwise.

We now consider the Coxeter elements corresponding to the above graphs. The characteristic polynomials can be computed as in [Ebeling$_4$], and one gets:

$$p^{(2)}(\Theta_{p_1,\ldots,p_h})(t) = (t-1)^2 \prod_{i=1}^{h} \frac{t^{p_i}-1}{t-1},$$

$$p^{(2)}(\tilde{\Theta}_{p_1,\ldots,p_h})(t) =$$

$$= (t^3-2t^2-2t+1) \prod_{i=1}^{h} \frac{t^{p_i}-1}{t-1} + t^2 \sum_{i=1}^{h} \frac{t^{p_i}-1}{t-1} \prod_{\substack{j=1 \\ j\neq i}}^{h} \frac{t^{p_j}-1}{t-1} .$$

The second formula was computed by K. Saito.

The first formula shows that the Coxeter element $\hat{c}^{(n)}$ corresponding to the graph $\Theta_{p_1,\ldots,p_h}$ is quasi-unipotent for all values of $p_1,\ldots,p_h$. However, it is only semi-simple for the values $(p_1,\ldots,p_h) = (2,3,6)$, $(2,4,4)$, $(3,3,3)$, $(2,2,2,2)$. To prove this, consider the lattice $\hat{L}^{(2)}$ determined by the intersection matrix $A^{(2)}$ corresponding to the graph $\Theta_{p_1,\ldots,p_h}$. Define $m_0 = \dim(\ker \hat{L}^{(2)})$. By Corollary 1.6.4 the number m_0 is equal to the dimension of the eigenspace of $\hat{c}^{(2)}$ corresponding to the eigenvalue 1. But $m_0 = 1$ except for the above values of $p_1,\ldots,p_h$, for which $m_0 = 2$, whereas the multiplicity of the eigenvalue 1 is equal to 2 according to the above formula for the characteristic polynomial. One should also compare the proof of Proposition 3.6.2.

On the contrary, the Coxeter element corresponding to the graph $\tilde{\Theta}_{p_1,\ldots,p_h}$ is only quasi-unipotent for certain finitely many values of $p_1,\ldots,p_h$. More information is given by the following theorem, which is due to K. Saito [Saito$_2$].

<u>Theorem 3.4.3</u> (K. Saito). <u>Let</u> $\hat{c} = \hat{c}^{(2)}$ <u>be the Coxeter element corresponding to the graph</u> $\tilde{\Theta}_{p_1,\ldots,p_h}$. <u>Then the following is true:</u>
<u>The operator</u> $\hat{c}$ <u>is quasi-unipotent if and only if</u> $(p_1,\ldots,p_h)$ <u>is one of the</u> h-<u>tuples of Table 3.4.2.</u> <u>The Coxeter element</u> $\hat{c}$ <u>is quasi-unipotent and semi-simple if and only if the corresponding</u> h-<u>tuple</u>

is not enclosed in square brackets. A quasi-unipotent $\hat{c}$ satisfies the following condition (*) if and only if the corresponding h-tuple is not enclosed in round brackets:

(*) An N-th primitive root of unity is an eigenvalue of $\hat{c}$, where N is the order of $\hat{c}$.

Table 3.4.2.

h = 3 :	[236] ,	237 ,	238 ,	239 ,	(23,10) ,	
	[244] ,	245 ,	246 ,	247 ,	(248) ,	
			255 ,	256 ,	(257) ,	(266) ,
	[333] ,	334 ,	335 ,	336 ,	(337) ,	
			344 ,	345 ,	(346) ,	(355) ,
				444 ,	(445) .	
h = 4 :	[2222] ,	2223 ,	2224 ,	2225 ,	(2226) ,	
			2233 ,	2234 ,	(2235) ,	(2244) ,
				2333 ,	(2334) ,	
					[(3333)].	
h = 5 :			22222 ,	22223 ,	[(22224)],	(22233) .
h = 6 :					(222222).	

We now consider the problem which of the graphs $\Theta_{p_1,\ldots,p_h}$ and $\tilde{\Theta}_{p_1,\ldots,p_h}$ occur as Dynkin diagrams of singularities. Up to now we have the following facts. Let us first consider the case h = 3. Then the graphs $\Theta_{p,q,r}$, for $\frac{1}{p} + \frac{1}{q} + \frac{1}{r} \leq 1$, are precisely the Dynkin diagrams of [Gabrielov$_2$] corresponding to strongly distinguished bases of vanishing cycles of the unimodal hypersurface singularities of type $T_{p,q,r}$ in Arnol'd's notation. (Here unimodal means unimodal with respect to right-equivalence .) The graphs $\tilde{\Theta}_{p,q,r}$ for the 14 values of Table 3.4.2 not enclosed in brackets are exactly the corresponding Dynkin diagrams of [Gabrielov$_2$] for the 14 exceptional unimodal hypersurface singularities.

For the case $h = 4$ the following is true: The graphs $\Theta_{2,2,r,s}$ are Dynkin diagrams corresponding to strongly distinguished bases of thimbles of the singularities $T^1_{2,r,2,s}$, which include the simple space curve singularities S_μ, $\mu \geq 5$, and T_μ, $\mu = 7,8.9$. Wall denotes these singularities by $P_{r,s}$ [Wall$_3$]. Their deformation theory has recently been investigated by K. Wirthmüller [Wirthmüller$_1$],[Wirthmüller$_2$]. Finally the $\tilde{\Theta}_{2,2,r,s}$ for the 5 quadruples (2,2,r,s) of Table 3.4.2 not enclosed in brackets are the corresponding Dynkin diagrams for the remaining 5 simple space curve singularities given by map-germs with regular 2-jets.

We shall give in Section 3.6 an interpretation of the graphs $\tilde{\Theta}_{p_1,\ldots,p_h}$ for $h = 3,4$, where $(p_1,\ldots,p_h)$ is one of the h-tuples of Table 3.4.2 enclosed in round, but not square, brackets.

3.5. Surfaces of complete intersection

We shall now study more precisely the case $n = 2$, hence the case of isolated singularities given by map-germs

$$f : (\mathbb{C}^4,0) \longrightarrow (\mathbb{C}^2,0)$$

with regular 2-jets j^2f. According to Sections 3.1 up to 3.3, it remains to deal with the cases with Segre symbols $\{4\}$ and $\{(3,1)\}$, which form the L- and the M-series respectively. In this section we compute Dynkin diagrams for these singularities, where the procedure is analogous to Section 3.2 and Section 3.3.

By [Wall$_4$] a mapping f with Segre symbol $\{4\}$ (L-series) can be reduced to the prenormal form

$$f(w,x,y,z) = (2xz+2wy, \sqrt{-1}y^2+2wx + R(w,y,z))$$

with $R(w,y,z) = a(z) + yb(z) + wd(z) + w^2e(z)$ and ord $a \geq 3$; ord b, ord $d \geq 2$; ord $e \geq 1$.

A mapping f with Segre symbol $\{(3,1)\}$ (M-series) can be reduced to the prenormal form

$$f(w,x,y,z) = (2xz+2wy, -2x(w+y)+T(w,y,z)+xh(z))$$

with $T(w,y,z) = a(z) + yb(z) + (y+w)d(z) + (y-w)^2e(z)$ and ord $a \geq 3$; ord b, ord d, ord $h \geq 2$; and ord $e \geq 1$. Here again some terms are superfluous, but are used in order to simplify some equations.

First we consider a suitable projection of the surface $X = f^{-1}(0)$. For this purpose, we eliminate in both cases the variable x from the equations $f_1(w,x,y,z) = 0$ and $f_2(w,x,y,z) = 0$.

Then one obtains a function $L_x f$ in the variables w,y,z, belonging to one of the series in Arnol'd's classification [Arnol'd$_2$]:

$$\text{L-series:}\quad (L_x f)(w,y,z) = -4w^2y+2\sqrt{-1}y^2z + 2zR(w,y,z) \quad \text{(S-series)}$$

$$\text{M-series:}\quad (L_x f)(w,y,z) = 4wy(w+y-h(z)) + 2zT(w,y,z) \quad \text{(U-series)}$$

Geometrically, this elimination corresponds to the following operation. The surface $X = f^{-1}(0)$ contains the x-axis. The equation $(L_x f)(w,y,z)=0$ is the equation of the projection of X on the (w,y,z) - space, where the x-axis is mapped onto the origin. We choose the coordinate function z as the linear function ζ. Then the polar curve $\Sigma_z(f_2)$ is the union of the x-axis with a residual curve $\tilde{\Sigma}$. The projection of $\tilde{\Sigma}$ on the (w,y,z) -space is just the polar curve $\Sigma_z(L_x f)$. Let

$$\Sigma_z(L_x f) = \bigcup_{i=1}^{\ell} \Sigma_i^L$$

be the decomposition of $\Sigma_z(L_x f)$ into irreducible components Σ_i^L, and let ρ_i^L be the corresponding invariants of Chapter 2.1. Let Σ_i, $i = 1,\dots,\ell$, be the component of $\tilde{\Sigma}$ projected onto Σ_i^L, and let ρ_i and ν_i be the corresponding invariants. Those components Σ_i^L which are not contained in the hyperplane $z = 0$ have a generalized Puiseux parametrization by [Maurer$_1$]

$$w = b_i z^{\tilde{\beta}_i} + \quad \text{higher terms,} \qquad b_i \in \mathbb{C},\ \tilde{\beta}_i \in \mathbb{Q},\ b_i \neq 0,$$

$$y = c_i z^{\tilde{\gamma}_i} + \quad \text{higher terms,} \qquad c_i \in \mathbb{C},\ \tilde{\gamma}_i \in \mathbb{Q},\ c_i \neq 0.$$

Inserting this parametrization into the function $L_x f$, one can determine the numbers ρ_i^L. It follows from the equation $f_1(w,x,y,z) = 0$ that the coordinate x can be written on Σ_i as a fractional power series in z as follows:

$$x = d_i z^{\tilde{\beta}_i+\tilde{\gamma}_i-1} + \text{higher terms,} \qquad d_i \in \mathbb{C},\ d_i \neq 0.$$

Inserting this parametrization of Σ_i into the function f_2 yields the numbers ρ_i. A comparison of these invariants shows that

$$\rho_i = \rho_i^L - 1 \quad \text{for } i \quad \text{with } \Sigma_i^L \not\subset \{z = 0\} .$$

In the case of the L-series one component of the polar curve $\Sigma_z(L_x f)$ is the line $\{w = z = 0\}$. This line is contained in the hyperplane $\{z = 0\}$. The fact that one component of the polar curve $\Sigma_z(L_x f)$ lies in the hyperplane $\{z = 0\}$ is equivalent to the fact that $(L_x f)|_{z=0}$ has a non-isolated singularity (cf. [Gabrielov$_3$]). The line $\Sigma_1 = \{w = z = 0\}$ contains two critical points of $L_x f - 2\varepsilon z$. For the remaining components Σ_i, $i > 1$, one computes

$$\sum_{i=2}^{\ell} \nu_i = 3.$$

Finally one has $\mu(f_1, f_2 - 2\varepsilon z) = 4$.

In the case of the M-series no component of the polar curve $\Sigma_z(L_x f)$ lies in the hyperplane $\{z = 0\}$. In this case we have

$$\sum_{i=1}^{\ell} \nu_i = 4, \quad \mu(f_1, f_2 - 2\varepsilon z) = 3.$$

In both cases one can then deduce from Proposition 2.1.4 and Proposition 2.1.5:

<u>Lemma 3.5.1.</u> <u>For a map-germ</u> f <u>belonging to the L- or M-series the following is true:</u> $\mu(f) = \mu(L_x f) - 1$.

One can also show in this way that the assignment $f \longmapsto L_x f$ again yields bijections between the μ-constant strata.

The mapping $g = (f_1, f_2 + z^2)$ defines in both cases a singularity of type $\tilde{D}_5$. We shall now study the behaviour of the critical values of $g_{\varepsilon,t} = f_2 + (z - \varepsilon)^2|_{X_t'}$ for $\varepsilon \longrightarrow 0$ and the corresponding Dynkin diagrams in detail.

(A) <u>L-series</u>

The critical values of $g_{\varepsilon,t}$ are given asymptotically (up to terms of order $o(\varepsilon^2)$) for $\varepsilon \longrightarrow 0$ and $t \longrightarrow 0$ as follows: They are the images of the solutions of the equation

$$z^5(z - \varepsilon)^3 = \sqrt{-1} \left(\tfrac{t}{4}\right)^4$$

under the mapping

$$z \longmapsto 4(z - \tfrac{5}{8}\varepsilon)^2 - \tfrac{9}{16}\varepsilon^2 \quad .$$

We denote these critical values by $s_i^{\varepsilon,t}$, $i = 1,\ldots,8$, where the order corresponds to the order of the system of paths $(\varphi_1^{\varepsilon_0},\ldots,\varphi_8^{\varepsilon_0})$, which we choose in order to connect the critical values $s_i^{\varepsilon_0,t}$ with the origin. Then the behaviour of the critical values for $\varepsilon = \varepsilon_0 \longrightarrow \varepsilon = 0$ is illustrated in Figure 3.5.1. Like Figure 3.3.1, this figure is based on a computer graphic. Compare the remarks there. The exact asymptotics of the three critical values $s_1^{\varepsilon,t}$, $s_4^{\varepsilon,t}$, and $s_8^{\varepsilon,t}$ depend on $R(w,y,z)$ and are given by Proposition 2.2.1.

$\varepsilon_0 = 0,1$

Figure 3.5.1

$t \sim 0,0024$

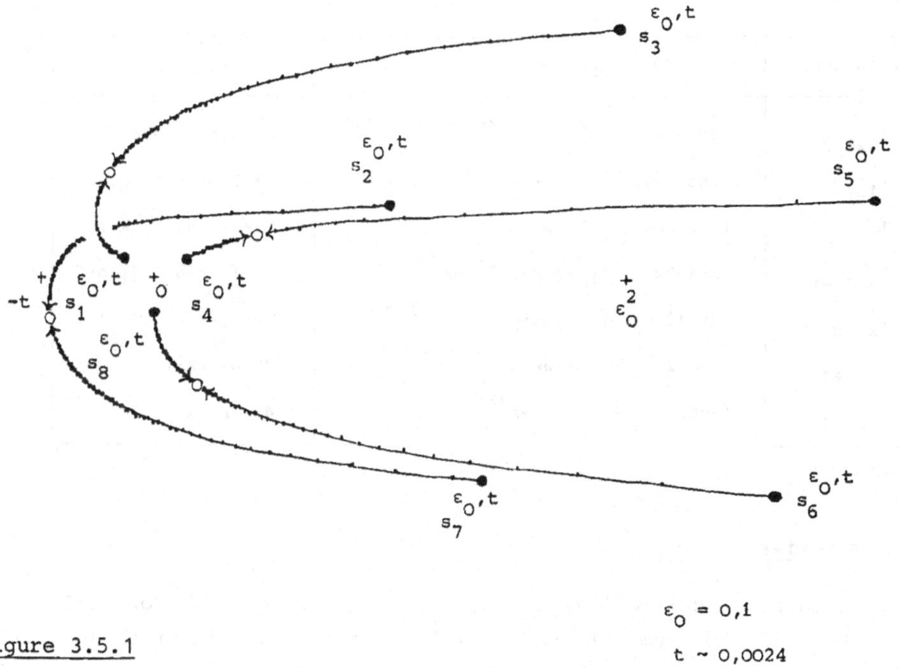

An analogous calculation as in Section 3.3 then yields a strongly distinguished basis $(e_j^r \mid 1 \le j \le 8,\ 1 \le r \le M_j)$ of thimbles for f which has the same properties as the corresponding bases for the $J^{(n-1)}$- and $K^{(n-1)}$-series, namely:

(i) The intersection matrix of $(e_1^1,\dots,e_8^1)$ coincides with the intersection matrix of $(\hat{\delta}_1',\dots,\hat{\delta}_8')$ for the singularity $\tilde{D}_5$ (Chapter 2.3).

(ii) The complete intersection matrix is related to the intersection matrix of $(e_1^1,\dots,e_8^1)$ by the formulas of Theorem 2.2.3.

The numbers M_j have the following values:

$$M_j = 1 \quad \text{for} \quad j \neq 2,4,7, \quad M_2 = M_4^L - 1,\ M_4 = M_2^L - 1,\ M_7 = M_3^L - 1;$$

where M_2^L, M_3^L, and M_4^L are the corresponding numbers for the S-series of [Gabrielov$_3$]. The individual values of M_2, M_4, and M_7 are tabulated in Table 3.5.1.

Table 3.5.1.

Notation	Mapping f	M_2,M_4,M_7
L_{12k-2}	$(2xz+2wy, \sqrt{-1}y^2+2wx+z^{4k-1})$	4k-2,4k-2,4k-2
L_{12k-1}	$(2xz+2wy, \sqrt{-1}y^2+2wx+wz^{3k-1})$	4k-2,4k-1,4k-2
$L_{k,i}$	$(2xz+2wy, \sqrt{-1}y^2+2wx+w^2z^{2k-1}+z^{4k+i})$	4k+i-1,4k-1,4k-1
$L^{\#}_{k,2q-1}$	$(2xz+2wy, \sqrt{-1}y^2+2wx+yz^{2k}+wz^{3k+q-1})$	4k-1,4k+q-1,4k+q-2
$L^{\#}_{k,2q}$	$(2xz+2wy, \sqrt{-1}y^2+2wx+yz^{2k}+w^2z^{2k+q-1})$	4k-1,4k+q-1,4k+q-1
L_{12k+3}	$(2xz+2wy, \sqrt{-1}y^2+2wx+wz^{3k})$	4k,4k,4k-1
L_{12k+4}	$(2xz+2wy, \sqrt{-1}y^2+2wx+z^{4k+1})$	4k,4k,4k

(B) M-series

In this case the critical values of $g_{\varepsilon,t}$ are given asymptotically for $\varepsilon,t \longrightarrow 0$ up to terms of order $o(\varepsilon^2)$ as follows: Six of these critical values are the images of the solutions of the equation

$$(z^2(z-\varepsilon) + \sqrt{-2}(\sqrt{t})^3)(z^2(z-\varepsilon) - \sqrt{-2}(\sqrt{t})^3) = 0$$

under the mapping

$$z \longmapsto -z^2 + \varepsilon^2 \quad .$$

This equation coincides up to a constant with the equation (*) of Section 3.3 (A) ($J^{(n-1)}$-series). In the same way the mapping only differs by a translation of z and by a constant from the mapping given there. Hence the qualitative behaviour of these six critical values is the same as that in Section 3.3 (A) and is given by Figure 3.3.1. In addition for the M-series, there are two further critical points, which have the critical value 0 up to order $o(\varepsilon^2)$. The exact asymptotics for those critical values which coincide with 0 up to order $o(\varepsilon^2)$ for $t = 0$ depend on $T(w,y,z)$ and $h(z)$ and are given by Proposition 2.2.1.

An analogous calculation as in Section 3.3 then yields a strongly distinguished basis $(\tilde{e}_j^r \mid 1 \le j \le 8,\ 1 \le r \le \tilde{M}_j)$ of thimbles for f with the following properties:

(i') The Dynkin diagram corresponding to $(\tilde{e}_1^1,\dots,\tilde{e}_8^1)$ is the graph of Figure 3.5.2.

(ii) The complete intersection matrix is related to the intersection matrix of $(\tilde{e}_1^1,\dots,\tilde{e}_8^1)$ by the formulas of Theorem 2.2.3.

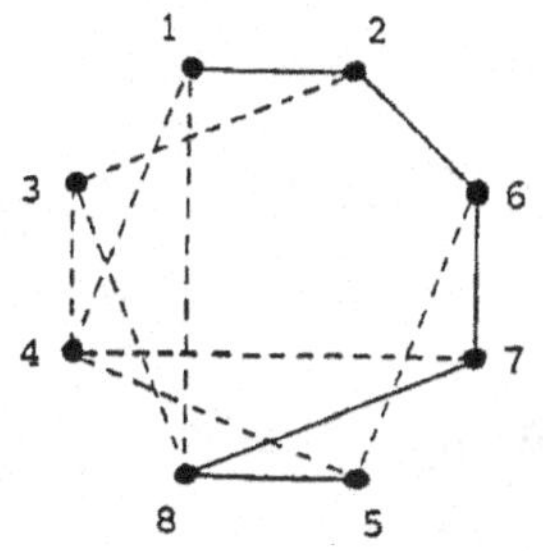

Figure 3.5.2

Here $\tilde{M}_j = 1$ for $j \neq 1,4,7,8$, and the individual values of $\tilde{M}_1$, $\tilde{M}_4$, $\tilde{M}_7$, and $\tilde{M}_8$ are given in Table 3.5.2.

Table 3.5.2.

Notation	Mapping f	$\tilde{M}_1, \tilde{M}_4, \tilde{M}_7, \tilde{M}_8$
M_{12k-1}	$(2xz+2wy, -2x(w+y)+z^{3k})$	3k-1, 3k-1, 3k-1, 3k-1
$M_{k,2q-1}$	$(2xz+2wy, -2x(w+y)+(w+y)z^{2k}+(w-y)^2 z^{k+q-1})$	3k, 3k+q-1, 3k-1, 3k+q-1
$M_{k,2q}$	$(2xz+2wy, -2x(w+y)+(w+y)z^{2k}+(w-y)z^{2k+q})$	3k, 3k+q, 3k-1, 3k+q-1
$M^{\#}_{k,2q-1}$	$(2xz+2wy, -2x(w+y)+yz^{2k}+(w-y)^2 z^{k+q-1})$	3k+q-1, 3k, 3k+q-1, 3k-1
$M^{\#}_{k,2q}$	$(2xz+2wy, -2x(w+y)+yz^{2k}+(w-y)z^{2k+q})$	3k+q, 3k, 3k+q-1, 3k-1
M_{12k+3}	$(2xz+2wy, -2x(w+y)+z^{3k+1})$	3k, 3k, 3k, 3k

In the cases M_{12k-1}, $M_{k,2q-1}$, $M_{k,2q}$, and M_{12k+3} the basis $(\tilde{e}^r_j)$ can be transformed into a strongly distinguished basis (e^r_j) of thimbles which has the same properties (i) and (ii) as the corresponding bases of the L-series and of the other series: The intersection matrix corresponding to $(e^1_1, \ldots, e^1_8)$ coincides with the intersection matrix of $(\hat{\delta}'_1, \ldots, \hat{\delta}'_8)$ of Chapter 2.3. Put $(\tilde{e}_{(r,j)}) = (\tilde{e}^r_j)$. Then one can obtain such a basis $(e^r_j \mid 1 \le j \le 8,\ 1 \le r \le M_j)$ from the basis $(\tilde{e}_{(r,j)})$ by the transformations

$$\beta_{(1,8)}, \beta_{(2,8)}, \ldots, \beta_{(\tilde{M}_1,8)}; \beta_{(1,7)}, \beta_{(1,6)}, \alpha_{(1,1)}, \beta_{(1,8)} .$$

For the numbers M_j one gets

$$M_j = 1 \quad \text{for} \quad j \neq 2,4,5,7, \quad (M_2, M_4, M_5, M_7) = (\tilde{M}_1, \tilde{M}_4, \tilde{M}_8, \tilde{M}_7).$$

If one applies the same transformations to the basis $(\tilde{e}^r_j)$ for $M^{\#}_{k,2q-1}$ and $M^{\#}_{k,2q}$, then condition (ii) is no longer satisfied for the transformed basis (e^r_j). Note in this context that there is an error in Table 1 of [Gabrielov$_3$] concerning $U_{k,2q}$, as the author has observed in [Zentralblatt für Mathematik 421.32011]: Also in this case one has to take a different Dynkin diagram for $f|_{z=0}$ as a basis.

3.6. Strange duality

In this section we consider some isolated complete intersection singularities of dimension 2, which are at the beginning of the hierarchy of all such singularities. After the simple hypersurface singularities one has the following classes of singularities (cf. e.g. [Looijenga$_3$, (7.23)]):

(a) The simply elliptic singularities $\tilde{E}_8$, $\tilde{E}_7$, $\tilde{E}_6$, $\tilde{D}_5$.

(b) The hyperbolic singularities $T_{p,q,r}$ and $T^2_{p,q,r,s}$. These are the two-dimensional cusp singularities of embedding dimension less than or equal to 4.

(c) The triangle singularities $D_{p,q,r}$ (in the notation of [Looijenga$_2$], [Looijenga$_3$], [Looijenga$_4$]) of embedding dimension less than or equal to 4. These are the 14 exceptional unimodal (with respect to right-equivalence) hypersurface singularities and the 8 singularities J'_9, J'_{10}, J'_{11}, K'_{10}, K'_{11}, L_{10}, L_{11}, and M_{11} of embedding dimension 4.

These singularities are unimodal with respect to K-equivalence. However, this list does not exhaust the class of two-dimensional K-unimodal singularities. We shall consider further K-unimodal singularities later.

The singularities of embedding dimension 3 of this list are precisely the unimodal hypersurface singularities in the sense of right-equivalence. We have already considered the Dynkin diagrams of Gabrielov for these singularities in Section 3.4.

For the above singularities of embedding dimension 4 we have already obtained the following Dynkin diagrams corresponding to strongly distinguished bases of thimbles. According to Chapter 2.3 the singularity $\tilde{D}_5$ has the Dynkin diagram $\Pi^2_{2,2,2,2}$. The singularity $T^2_{p,q,r,s}$ has the Dynkin diagram $\Pi^2_{p,q,r,s}$ for $p = r = 2$ and for $(p,q,r,s) = (2,2,3,3)$ according to Section 3.2 and to Chapter 2.5 respectively. We conjecture that this is more generally true for any $p,q,r,s \geq 2$. The triangle singularities belonging to the J'- and K'-series have Dynkin diagrams of type $\tilde{\Pi}^2_{p,q,r,s}$ according to Proposition 3.3.4. The graph $\tilde{\Pi}^2_{p,q,r,s}$ is shown once more in Figure 3.6.1. The graph $\Pi^2_{p,q,r,s}$ is the subgraph without the vertex with number $\rho - 2$. One has

$$\rho = p + q + r + s + 1.$$

We assume throughout that the quadruple (p,q,r,s) is ordered in such a way that one has

$$2 \leq p \leq r, \quad 2 \leq q \leq s\ , \quad \text{and} \quad p < q \quad \text{or} \quad p = q,\ r \leq s.$$

<u>Proposition 3.6.1. The eight triangle singularities in $\mathbb{C}^4$ have strongly distinguished bases of thimbles with Dynkin diagrams $\tilde{\Pi}^2_{p,q,r,s}$, where the values of p,q,r,s are given in Table 3.6.1.</u>

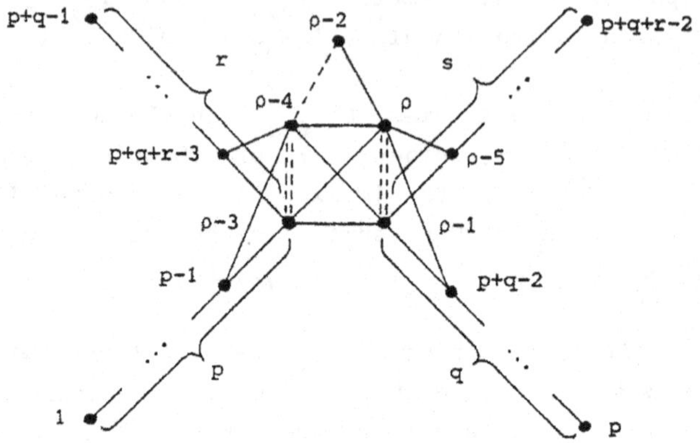

Figure 3.6.1: The graphs $\Pi^2_{p,q,r,s}$ and $\tilde{\Pi}^2_{p,q,r,s}$

Table 3.6.1.

Notation of Wall	Notation of Looijenga	p q r s	N	Dual singularity
J'_9	$D_{2,3,10}$	2 2 2 3	18	$E_{3,-1}$
J'_{10}	$D_{2,4,8}$	2 2 2 4	14	$Z_{1,-1}$
J'_{11}	$D_{3,3,7}$	2 2 2 5	12	$Q_{2,-1}$
K'_{10}	$D_{2,6,6}$	2 3 2 3	12	$W_{1,-1}$
K'_{11}	$D_{3,5,5}$	2 3 2 4	10	$S_{1,-1}$
L_{10}	$D_{2,5,7}$	2 2 3 3	12	$W^{\#}_{1,-1}$
L_{11}	$D_{3,4,6}$	2 2 3 4	10	$S^{\#}_{1,-1}$
M_{11}	$D_{4,4,5}$	2 3 3 3	9	$U_{1,-1}$

Proof. We show the assertion for the singularities J_9', K_{10}', L_{10}, and M_{11}. For the first two singularities the transformations are also already contained in the proof of Proposition 3.3.4. As in that proof, we only indicate those transformations which turn the Dynkin diagrams computed in Sections 3.3 and 3.5 into graphs of type $\tilde{\Pi}^2_{p,q,r,s}$ with possibly different numbering and orientation of the corresponding basis elements. In each case let $(e_1,\ldots,e_m) = (e_j^r \mid 1 \le j \le 8, 1 \le r \le M_j)$ be the strongly distinguished basis of thimbles of Sections 3.3 and 3.5, where the intersection matrix of $(e_1,\ldots,e_8) = (e_1^1,\ldots,e_8^1)$ coincides with the intersection matrix of $(\hat{\delta}_1',\ldots,\hat{\delta}_8')$ of Chapter 2.3. Here the numbers M_j are equal to 1 or 2, and M_j is equal to 2 for the following numbers:

Series	J'	K'	L	M
$M_j = 2$	M_2, M_7	M_2, M_4, M_5	M_2, M_4, M_7	M_2, M_4, M_5, M_7

We apply the following transformations to the above basis:

$$\alpha_2,\alpha_3,\alpha_4;\alpha_1,\alpha_2,\alpha_3,\alpha_4;$$

$$(L_{10}, M_{11} : \alpha_9),\ (M_{11} : \alpha_{10}),$$

$$(L_{10}, M_{11} : \alpha_8,\alpha_7,\ldots,\alpha_1),\ (M_{11} : \alpha_9,\alpha_8,\ldots,\alpha_1),$$

$$\alpha_{8+i}\ (*),\ \alpha_{7+i}\ (*),\ \alpha_{6+i}\ (*);\ \alpha_{8+i},\alpha_{7+i};$$

$$\beta_{2+i},\beta_{3+i};\varkappa_{7+i},\varkappa_{8+i},\varkappa_{9+i};$$

where $i = 0$ for J_9', K_{10}', $i = 1$ for L_{10}, and $i = 2$ for M_{11}. Up to the transformations enclosed in parentheses and those denoted by (*), these are transformations of the underlying Dynkin diagram of $\tilde{D}_5$ and coincide with the transformations (B) and (C) of Chapter 2.3 up to transformations which only alter the numbering. This proves Proposition 3.6.1.

We now examine the graphs $\Pi^2_{p,q,r,s}$ and $\tilde{\Pi}^2_{p,q,r,s}$ more closely. We use the notations of Section 3.4. It turns out that there is a close relation between these graphs and the graphs $\Theta_{p,q,r,s} = \Pi^1_{p,q,r,s}$ and $\tilde{\Theta}_{p,q,r,s} = \tilde{\Pi}^1_{p,q,r,s}$ of Section 3.4.

<u>Proposition 3.6.2.</u> <u>Let</u> D^n <u>be the graph</u> $\Pi^n_{p,q,r,s}$ <u>or</u> $\tilde{\Pi}^n_{p,q,r,s}$ <u>for</u> $n = 1,2$. <u>Here we assume</u> $3 \le s$, <u>if</u> $D^n = \tilde{\Pi}^n_{p,q,r,s}$. <u>Then the following is true:</u>

(i) $p^{(2)}(D^2)(t) = (t-1)^2 p^{(2)}(D^1)(t)$

(ii) <u>The Jordan normal form of the Coxeter element</u> $\hat{c}^{(2)}(D^2)$ <u>is</u>

obtained from the Jordan normal form of the Coxeter element $\hat{c}^{(2)}(D^1)$ by adding a Jordan block of the form

$$\begin{pmatrix} 1 & 1 \\ 0 & 1 \end{pmatrix}.$$

The eigenspace corresponding to the eigenvalue 1 of this block is generated by the vector

$$\lambda_\rho^{(2)} - \lambda_{\rho-1}^{(2)} + \lambda_{\rho-4}^{(2)} - \lambda_{\rho-3}^{(2)} .$$

Proof. (i) By the transformations

$$\beta_{\rho-1}, \beta_{\rho-2}, \beta_{\rho-3}, \beta_{\rho-3}, \beta_{\rho-2}, \beta_{\rho-1}, \kappa_{\rho-2} ,$$

the basis $(\lambda_1^{(2)},\ldots,\lambda_\rho^{(2)})$ corresponding to the graph $\tilde{\Pi}^2_{p,q,r,s}$ is turned into a basis $(\lambda_1',\ldots,\lambda_\rho')$ satisfying the following conditions

(a) $\langle\lambda_{\rho-1}',\lambda_\rho'\rangle = -2$, $\langle\lambda_i',\lambda_{\rho-1}'\rangle = \langle\lambda_i',\lambda_\rho'\rangle$ for $1 \le i \le \rho-2$.

(b) The Dynkin diagram corresponding to $(\lambda_1',\ldots,\lambda_{\rho-2}')$ is the graph $\tilde{\Pi}^1_{p,q,r,s}$.

The analogous facts are valid for the basis $(\lambda_1^{(2)},\ldots,\lambda_{\rho-3}^{(2)},\lambda_{\rho-1}^{(2)},\lambda_\rho^{(2)})$ corresponding to the graph $\Pi^2_{p,q,r,s}$.

Then the matrix $\hat{C}^{(2)}(D^2)$ of $\hat{c}^{(2)}(D^2)$ with respect to the basis $(e_1,\ldots,e_\rho) := (\lambda_1',\ldots,\lambda_{\rho-2}',\lambda_\rho'-\lambda_{\rho-1}',-\frac{1}{2}\lambda_\rho')$ has the following shape (in the case $D^2 = \Pi^2_{p,q,r,s}$ one has to omit the vector $\lambda_{\rho-2}'$):

$$\hat{C}^{(2)}(D^2) = \left(\begin{array}{ccc|c|c} & & & 0 & c_{1\rho} \\ & \hat{C}^{(2)}(D^1) & & \vdots & \vdots \\ & & & 0 & c_{\rho-2,\rho} \\ \hline * & \cdots & * & 1 & 1 \\ \hline 0 & \cdots & 0 & 0 & 1 \end{array}\right)$$

This implies assertion (i).

(ii) In the case $D^2 = \tilde{\Pi}^2_{p,q,r,s}$ one has

$$c_{i\rho} = \begin{cases} \frac{1}{2} & \text{for } i = 1,\ldots,p-1; p+q-1,\ldots,p+q+r-3; \rho-3, \rho-2; \\ 0 & \text{for } 1 \le i \le \rho-2 \quad \text{otherwise.} \end{cases}$$

We set

$$e_i' = \begin{cases} e_i + \frac{1}{2(q-1)} e_{\rho-1} & \text{for } i = p,\dots,p+q-2 ; \\ e_i + \frac{1}{2(s-1)} e_{\rho-1} & \text{for } i = p+q+r-2,\dots,\rho-5 ; \\ e_i - e_{\rho-1} & \text{for } i = \rho-2 ; \\ e_i & \text{otherwise.} \end{cases}$$

We abbreviate $\hat{c} = \hat{c}^{(2)}(D^2)$. Then one has

$$\hat{c}(e_i') \subset \langle e_1',\dots,e_{\rho-2}'\rangle \quad \text{for } i = 1,\dots,\rho-2 ,$$

$$\hat{c}(e_{\rho-1}') = e_{\rho-1}' ,$$

$$\hat{c}(e_\rho') = e_\rho' + \frac{3}{2} e_{\rho-1}' + \sum_{i=1}^{\rho-2} c_{i\rho} e_i' .$$

We set $\hat{c}' = \hat{c}^{(2)}(D^1)$. Provided that $3 \leq s$, the matrix $A^{(2)}(D^1)$ is non-degenerate. Therefore $\hat{c}'$ has not the eigenvalue 1. Let

$$\xi = (\hat{c}'-\mathrm{id})^{-1}\left(\sum_{i=1}^{\rho-2} c_{i\rho} e_i'\right) ,$$

$$e_i^* = \begin{cases} e_\rho' - \xi & \text{for } i = \rho , \\ e_i' & \text{otherwise.} \end{cases}$$

Then

$$\hat{c}(e_\rho^*) = e_\rho^* + \frac{3}{2} e_{\rho-1}^* ,$$

$$\hat{c}(e_{\rho-1}^*) = e_{\rho-1}^* ,$$

$$\hat{c}(e_i^*) = \hat{c}'(e_i^*) \subset \langle e_1^*,\dots,e_{\rho-2}^*\rangle \quad \text{for } i = 1,\dots,\rho-2 .$$

This implies the second assertion in the case $D^2 = \tilde{\Pi}^2_{p,q,r,s}$.

In the case $D^2 = \Pi^2_{p,q,r,s}$ one has

$$c_{i\rho} = \begin{cases} -\frac{1}{2} & \text{for } i = p,\dots,p+q-2; p+q+r-2,\dots,\rho-5; \rho-4 ; \\ \frac{1}{2} & \text{for } i = \rho-3 ; \\ 0 & \text{for } 1 \leq i \leq \rho-3 \quad \text{otherwise.} \end{cases}$$

We set in this case

$$e_i^* = \begin{cases} e_i + \frac{1}{2(q-1)} e_{\rho-1} & \text{for } i = p,\dots,p+q-2 ; \\ e_i + \frac{1}{2(s-1)} e_{\rho-1} & \text{for } i = p+q+r-2,\dots,\rho-5; \\ e_i & \text{for } i \neq \rho \text{ otherwise} ; \end{cases}$$

$$e_\rho^* = e_\rho + \sum_{i=p}^{p+q-2} \frac{i}{2q} e_i^* + \sum_{i=p+q+r-2}^{\rho-5} \frac{i}{2s} e_i^* + \frac{1}{2}\left(\frac{1}{q} + \frac{1}{s} - 1\right)e_{\rho-4}^* - \frac{1}{2}\left(\frac{1}{q} + \frac{1}{s} - 1\right)e_{\rho-3}^* .$$

Then one has the same relations for $\hat{c}(e_i^*)$ as above. This completes the proof of Proposition 3.6.2

Using Proposition 3.6.2, we can apply the results of Section 3.4 to the graphs considered here. Let $(\lambda_1^{(2)},\dots,\lambda_{\rho-3}^{(2)},(\lambda_{\rho-2}^{(2)},)\lambda_{\rho-1}^{(2)},\lambda_\rho^{(2)})$ be a basis corresponding to the graph $D2 = \tilde{\Pi}^2_{p,q,r,s}$ or $D^2 = \Pi^2_{p,q,r,s}$ with intersection matrix $A^{(2)}(D^2)$, let $\hat{L}$ be the lattice spanned by this basis, and let $\hat{c}:\hat{L} \longrightarrow \hat{L}$ be the corresponding Coxeter element. Let L' be a primitive $\mathbb{Z}$-submodule of $\ker(\hat{c} - \mathrm{id})$ of rank 1. We set

$$L = \hat{L}/L' .$$

Let $c:L \longrightarrow L$ be the operator on the lattice L induced by the Coxeter element $\hat{c}$.

In the case $D^2 = \Pi^2_{p,q,r,s}$, it then follows from Section 3.4 and Proposition 3.6.2 that c is quasi-unipotent for all values $p,q,r,s \geq 2$, but not semi-simple except possibly for $(p,q,r,s) = (2,2,2,2)$.

In the cases $D^2 = \Pi^2_{2222}$ and $D^2 = \tilde{\Pi}^2_{p,q,r,s}$ we define

$$L' = \mathbb{Z} \cdot (\lambda_\rho^{(2)} - \lambda_{\rho-1}^{(2)} + \lambda_{\rho-4}^{(2)} - \lambda_{\rho-3}^{(2)}) .$$

Then it follows for Π^2_{2222} that c is semi-simple. In the remaining cases we obtain from Theorem 3.4.3 and Proposition 3.6.2:

<u>Corollary 3.6.3.</u> <u>The operator</u> c <u>corresponding to the graph</u> $\tilde{\Pi}^2_{p,q,r,s}$ <u>is quasi-unipotent if and only if</u> (p,q,r,s) <u>is one of the following quadruples:</u>

[2222], 2223 , 2224 , 2225 , (2226) ,

2233 , 2234 , (2235) , (2244) ,

2323 , 2324 , (2325) , (2424) ,

2333 , (2334) , (2343) ,

[(3333)].

It is additionally semi-simple if and only if the corresponding quadruple is not enclosed in square brackets. It is quasi-unipotent and satisfies condition (*) of Theorem 3.4.3 if and only if the corresponding quadruple is not enclosed in round brackets.

The eight graphs corresponding to the eight quadruples not enclosed in brackets are exactly the Dynkin diagrams of Proposition 3.6.1 for the eight triangle singularities in $\mathbb{C}^4$. They have quasi-homogeneous equations. The same is true for the singularity $\tilde{D}_5$ with the Dynkin diagram Π^2_{2222}. By Proposition 1.6.6 and Proposition 3.6.2, the module H' coincides with the above module L' for each of these singularities, and the operator c can be identified with the monodromy operator of the corresponding singularity. In the same way, one of the operators c (depending on L') for a graph $\Pi^2_{p,q,r,s}$ corresponding to a singularity $T^2_{p,q,r,s}$ can be identified with the monodromy operator of the singularity by (1.6.2).

From Proposition 3.6.2 and from our calculation of Dynkin diagrams, one can deduce the following surprising result.

Corollary 3.6.4. Let (X,Y) *be one of the following pairs of singularities:* (T^2_{2323}, T^2_{2233}), (K'_{10}, L_{10}), *or* (K'_{11}, L_{11}). *We denote the Milnor lattices of* X *and* Y *by* H_X *and* H_Y *respectively. Then the following is true:* $H_X \otimes \mathbb{Q} \cong H_Y \otimes \mathbb{Q} =: V$ *(as vector spaces, but not as quadratic spaces). The corresponding monodromy operators* $c_X : V \longrightarrow V$ *and* $c_Y : V \longrightarrow V$ *are conjugate over* $\mathbb{Q}$.

Remark 3.6.5. The singularities of the J'- and K'-series can be obtained from the space curve singularities of the J- and K-series by "suspension". Here suspension means the following operation (cf. [Pickl$_1$]): To a mapping $f : \mathbb{C}^3 \longrightarrow \mathbb{C}^2$ with

$f(x,y,z) = (f_1(x,y,z), f_2(x,y,z))$, we assign the mapping $f^e : \mathbb{C}^4 \longrightarrow \mathbb{C}^2$ with $f^e(w,x,y,z) = (f_1(x,y,z) + w^2,\ f_2(x,y,z))$. Analogously, one obtains the cusp singularities $T^2_{2,q,2,s}$ by suspension from the space curve singularities $T^1_{2,q,2,s}$. A comparison between the results of Section 3.4 and the above results shows that the following is true for the singularities of Table 3.4.1: If f has a Dynkin diagram $\Theta_{p,q,r,s}$ (respectively $\tilde{\Theta}_{p,q,r,s}$), then the suspension f^e has a Dynkin diagram $\Pi^2_{p,q,r,s}$ (respectively $\tilde{\Pi}^2_{p,q,r,s}$). The relation between the corresponding relative monodromy operators is given by Proposition 3.6.2.

There exists a strange duality among the 14 triangle singularities on hypersurfaces in $\mathbb{C}^3$, which was observed by Arnol'd in [Arnol'd$_2$]. In [Ebeling-Wall], we consider an extension of this duality to a class of singularities embracing in particular the triangle singularities of embedding dimension 4. In addition one has to take the following class of K-unimodal surface singularities, which follows upon the triangle singularities in the classification, into account:

(d) The Kodaira singularities of type $I^*_i(k_1,k_2,k_3,k_4)$ and of grade $D = \sum k_i \leq 4$ (for definitions and notations see [Ebeling-Wall]). This defines for a fixed quadruple (k_1,k_2,k_3,k_4) a series of singularities indexed by $i \geq 0$, which begins with the quadrangle singularity $D_{k_1+2,k_2+2,k_3+2,k_4+2}$ in Looijenga's notation [Looijenga$_4$] for $i = 0$. These are the eight bimodal (with respect to right-equivalence) series of hypersurface singularities and the eight series $J'_{1,i}$, $K'_{1,i}$, $K^\flat_{1,i} = K^{(1),\#}_{1,i}$, $L_{1,i}$, $L^\#_{1,i}$, $M_{1,i}$, $M^\#_{1,i}$ and $I_{1,i}$ of singularities of embedding dimension 4. Here the series $I_{1,i}$ is the only one of these series which is not covered by our calculations, since the singularities of this series are not given by map-germs with regular 2-jets.

The extension of Arnol'd's strange duality considered in [Ebeling-Wall] is a duality between the eight triangle singularities in $\mathbb{C}^4$ and the eight bimodal series of hypersurface singularities, and a duality of the eight series of Kodaira singularities of type I^*_i in $\mathbb{C}^4$ among each other. Here a series is in each case represented by a "virtual" singularity obtained by substituting $i = -1$. It is represented by a Dynkin diagram which one obtains by substituting $i = -1$, respectively $q = 0$ in the index $2q - 1$, in Table 1 of [Gabrielov$_3$] and in the tables of Sections 3.3 and 3.5. These Dynkin diagrams do not correspond to a genuine isolated complete intersection singularity of dimension 2, as can be shown. These Dynkin diagrams are strongly equivalent to graphs of type $\tilde{\mathcal{T}}_{p,q,r}$ or $\tilde{\Pi}^2_{p,q,r,s}$. For the eight bimodal series of hypersurface singularities one obtains in this way the eight graphs $\tilde{\mathcal{T}}_{p,q,r}$ where (p,q,r) is one of the eight triples of Table 3.4.2 for $h = 3$ enclosed in round brackets. For the corresponding seven series of singularities in $\mathbb{C}^4$ (where we have excluded the series $I_{1,i}$) one gets in this way the seven graphs $\tilde{\Pi}^2_{p,q,r,s}$ of Corollary 3.6.3 where (p,q,r,s) is one of the seven quadruples enclosed in round brackets, but not enclosed in square brackets.

The assignment of the triples (p,q,r) and the quadruples (p,q,r,s) to the various series is given in Table 3.6.1 and Table 3.6.2 respectively. In Table 3.6.1, we have indicated for each of the eight triangle singularities in $\mathbb{C}^4$ the dual singularity in the above sense. The corresponding numbers p,q,r for the dual singularity (series) coincide with the Dolgachev numbers of the triangle singularity, hence with the indices p,q,r of Looijenga's notation $D_{p,q,r}$ of the corresponding triangle singularity. Moreover, we have indicated in Table 3.6.1 the Coxeter number N of the triangle singularity, i.e. the order of the corresponding monodromy operator c of the singularity. It turns out that this number coincides with the degree of a quasi-homogeneous equation for the quadrangle singularity (of type I_0^*) of the corresponding dual series. Vice versa, the order of the Coxeter element corresponding to the graph $\tilde{\mathcal{T}}_{p,q,r}$ can be identified with the least common multiple of the degrees of quasi-homogeneous equations of the dual triangle singularity (for the latter see [Ebeling-Wall, (5.2)]).

Table 3.6.2.

"Sing."	$b_1b_2b_3b_4$	p q r s	N	$(w_1,w_2,w_3,w_4;d_1,d_2)$	dual to
$J'_{1,-1}$	2 2 2 6	2 2 2 6	40	(2,4,5,6;8,10)	$J'_{1,-1}$
$L^{\#}_{1,-1}$	2 2 3 5	2 2 3 5	56	(2,3,4,5;7,8)	$L^{\#}_{1,-1}$
$L_{1,-1}$	2 3 2 5	2 2 4 4	24	(2,3,4,5;7,8)	$K^{\flat}_{1,-1}$
$K^{\flat}_{1,-1}$	2 2 4 4	2 3 2 5	56	(2,3,4,4;6,8)	$L_{1,-1}$
$K'_{1,-1}$	2 4 2 4	2 4 2 4	24	(2,3,4,4;6,8)	$K'_{1,-1}$
$M^{\#}_{1,-1}$	2 3 3 4	2 3 3 4	42	(2,3,3,4;6,7)	$M^{\#}_{1,-1}$
$M_{1,-1}$	2 3 4 3	2 3 4 3	42	(2,3,3,4;6,7)	$M_{1,-1}$

In Table 3.6.2 we have given the following informations for each of the seven series $I_i^*(k_1,k_2,k_3,k_4)$ of Kodaira singularities in $\mathbb{C}^4$ with the series $I_{1,i}$ excluded:

- the notation (of Wall) for the "virtual" singularity (i = -1),
- the numbers $b_i = k_i + 2$, $i = 1,\ldots,4$,
- the numbers p,q,r,s of the corresponding Dynkin diagram of type $\tilde{\Pi}^2_{p,q,r,s}$
- the Coxeter number N, i.e. the order of the operator c corresponding to $\tilde{\Pi}^2_{p,q,r,s}$,
- the quasi-homogeneous type $(w_1,w_2,w_3,w_4;d_1,d_2)$ of quasi-homogeneous equations for the quadrangle singularity $I_0^*(k_1,k_2,k_3,k_4)$ of the corresponding series,
- the dual singularity.

One again observes that the Coxeter number N coincides with the least common multiple of the degrees of the quadrangle singularity (i = 0) of the corresponding dual series I_i^*.

We conclude by mentioning that one can also define a "virtual" singularity for i = -1 for the series $J_{1,i}$, $K_{1,i}$, and $K^{\#}_{1,i}$ of space curve singularities of Section 3.3. The corresponding Dynkin diagrams of Tables 3.3.1 and 3.3.2 with i = -1 are strongly equivalent to the graphs $\tilde{\Theta}_{2226}$, $\tilde{\Theta}_{2244}$, and $\tilde{\Theta}_{2235}$ respectivley. The corresponding quadruples (p,q,r,s) are exactly the quadruples of the form (2,2,r,s) of Table 3.4.2 for h = 4 enclosed in round brackets.

4. MILNOR LATTICES AND MONODROMY GROUPS

4.1 A description of the monodromy groups and vanishing lattices

In this chapter we give a description of the monodromy group Γ, of the relative monodromy group $\hat{\Gamma}$, of the set of vanishing cycles Δ, and of the set of thimbles $\hat{\Delta}$ of an even-dimensional isolated complete intersection singularity.

Let (X,x) be an isolated complete intersection singularity given by a map-germ $f:(\mathbb{C}^{n+k},0) \longrightarrow (\mathbb{C}^k,0)$. Let

$$0 \longrightarrow H' \longrightarrow \hat{H} \longrightarrow H \longrightarrow 0$$

be the corresponding fundamental exact sequence of lattices of Chapter 1.1. Let $\mathrm{Aut}\,(H)$ be the automorphism group of the integral (symmetric or skew-symmetric) lattice H, and $\Gamma \subset \mathrm{Aut}\,(H)$ be the monodromy group. The non-degenerate lattice $H/\ker H$ corresponding to H will be denoted by $\bar{H}$. We denote the image of Γ under the mapping $\mathrm{Aut}\,(H) \longrightarrow \mathrm{Aut}\,(\bar{H})$ by Γ_s and the kernel of the mapping $\Gamma \longrightarrow \Gamma_s$ by Γ_u. The subgroup Γ_s is called the <u>simple</u>, the subgroup Γ_u the <u>unipotent</u> part of the monodromy group. The unipotent part Γ_u of the monodromy group is abelian and contained in the subgroup $\ker\,(H) \otimes \bar{H}$ of $\mathrm{Aut}\,(H)$ [Looijenga$_3$, (7.11)]. Here we regard $\ker\,(H) \otimes \bar{H}$ as a subgroup of $\mathrm{Aut}\,(H)$, associating the following mapping with $v \otimes w \in \ker\,(H) \otimes \bar{H}$:

$$(v \otimes w)(y) := y + \langle y,w\rangle v \quad \text{for} \quad y \in H.$$

Now let L be an arbitrary even symmetric lattice with bilinear form $\langle\,,\rangle$. Let $\varepsilon = \pm 1$. To a minimal vector $v \in L$ of square length $\langle v,v\rangle = 2\varepsilon$ corresponds a reflection s_v defined by

$$s_v(y) = y - \frac{2\langle y,v\rangle}{\langle v,v\rangle} v \quad \text{for} \quad y \in L.$$

We denote by $R_\varepsilon(L)$ the subgroup of the group $O(L)$ of units of L, which is generated by all reflections s_v corresponding to minimal vectors v of square length $\langle v,v\rangle = 2\varepsilon$ (see Chapter 5.1).

We define a homomorphism

$$\sigma_\varepsilon : O(L) \longrightarrow \{+1,-1\}$$

(real spinor norm) as follows. Let $g \in O(L)$ and $\bar{g}$ be the induced element in $O(\bar{L}_{\mathbb{R}})$. Here $\bar{L}_{\mathbb{R}} = (L/\ker L) \otimes \mathbb{R}$. Then one can write $\bar{g}$ in $O(\bar{L}_{\mathbb{R}})$ as a product of reflections $\bar{g} = s_{v_1} \circ \ldots \circ s_{v_r}$, where now the vectors v_i may have arbitrary length. We define

$$\sigma_\varepsilon(g) := \begin{cases} +1 & \text{if } \varepsilon\langle v_i, v_i\rangle < 0 \text{ for an even number of indices,} \\ -1 & \text{otherwise.} \end{cases}$$

The subgroup of $O(L)$ consisting of all units g with $\sigma_\varepsilon(g) = 1$ is denoted by $O'_\varepsilon(L)$.

This subgroup can also be characterized in the following way. Let t_+, respectively t_-, be the dimension of a maximal positive, respectively negative, definite subspace of $L_{\mathbb{R}} = L \otimes \mathbb{R}$, and let t_0 be the rank of ker L. The set of all oriented maximal ε-definite subspaces of $L_{\mathbb{R}}$ forms an open subset of the Grassmannian $G^{or}_{t_\varepsilon}(L_{\mathbb{R}})$. It has two connected components, if L is indefinite. Then $g \in O'_\varepsilon(L)$ if and only if g leaves each component of this set invariant (cf. [Looijenga$_3$, (7.12)]).

We denote by $L^\#$ the dual lattice $\mathrm{Hom}(L,\mathbb{Z})$, and by $j : L \longrightarrow L^\#$ the natural homomorphism. Let τ be the canonical homomorphism

$$\tau : O(L) \longrightarrow \mathrm{Aut}\,(L^\#/j(L)).$$

Let $O^*_\varepsilon(L)$ be the subgroup of $O(L)$ consisting of all units $g \in O(L)$ with $\sigma_\varepsilon(g) = 1$ and $\tau(g) = 1$ (cf. Chapter 5.1). One has

$$R_\varepsilon(L) \subset O^*_\varepsilon(L).$$

Now let $L = H$, n be even, and $\varepsilon = (-1)^{n/2}$. The triple (t_-, t_0, t_+) is denoted by (μ_-, μ_0, μ_+) in this case. We call a singularity (and a lattice H) <u>parabolic</u>, if $\mu_{-\varepsilon} = 0$ and $\mu_0 > 0$, and <u>hyperbolic</u>, if $\mu_{-\varepsilon} = 1$.

With these definitions we can formulate the following results.

<u>Theorem 4.1.1.</u> <u>Let</u> (X,x) <u>be an isolated complete intersection singularity of even dimension</u> n. <u>Then the following is true:</u>

(i) The singularity (X,x) has an ε-definite intersection form if and only if (X,x) is a simple hypersurface singularity.

(ii) The singularity (X,x) is parabolic if and only if (X,x) is of type $\tilde{E}_6$, $\tilde{E}_7$, $\tilde{E}_8$ or $\tilde{D}_{n+3}$.

(iii) The singularity (X,x) is hyperbolic if and only if (X,x) is of type

$$T_{p,q,r}(p=2,\ q=3,\ 7\leq r;\ p=2,\ 4\leq q\leq r,\ 5\leq r;\ 3\leq p\leq q\leq r, 4\leq r),$$

$$T^2_{p,q,r,s}\ (2\leq p\leq r,\ 2\leq q\leq s,\ r\leq s, 3\leq s),\ \text{or}\ \ T^n_{2,q,2,s}(2\leq q\leq s,\ 3\leq s)\ .$$

Theorem 4.1.2. Let (X,x) be an isolated complete intersection singularity of even dimension n, which is not hyperbolic or hyperbolic of type T_{237}, T_{245}, T_{334}, T^2_{2223}, or T^4_{2223}. Let $\varepsilon = (-1)^{n/2}$. Then the following is true:

(i) The monodromy group Γ (relative monodromy group $\hat{\Gamma}$) can be described as follows:

$$\Gamma = R_\varepsilon(H) = O^*_\varepsilon(H)\ ,$$
$$\hat{\Gamma} = R_\varepsilon(\hat{H}) = O^*_\varepsilon(\hat{H})\ ,$$

(ii) The short exact sequence

$$1 \longrightarrow \Gamma_u \longrightarrow \Gamma \longrightarrow \Gamma_s \longrightarrow 1$$

splits (non-canonically) and is equal to the following (non-canonically) split short exact sequence

$$1 \longrightarrow \ker(H) \otimes \bar{H} \longrightarrow O^*_\varepsilon(H) \rightleftarrows O^*_\varepsilon(\bar{H}) \longrightarrow 1.$$

In particular the subgroup Γ_s is of finite index in $O(\bar{H})$, and hence arithmetic.

(iii) Assume that (X,x) is not an ordinary double point (i.e. not of type A_1). Then the set of vanishing cycles Δ (set of thimbles $\hat{\Delta}$) can be described as follows:

$$\Delta = \{v \in H \mid \langle v,v\rangle = 2\varepsilon \quad \text{and} \quad \langle v,H\rangle = \mathbb{Z}\}\ ,$$
$$\hat{\Delta} = \{v \in \hat{H} \mid \langle v,v\rangle = 2\varepsilon \quad \text{and} \quad \langle v,\hat{H}\rangle = \mathbb{Z}\}\ .$$

In particular all the invariants depend only on the Milnor lattice H, respectively on the lattice $\hat{H}$, except in the excluded hyperbolic cases.

Theorem 4.1.3. *For all isolated complete intersection singularities of even dimension the following is true:*

(i) *The unipotent part* Γ_u *of the monodromy group coincides with* $\ker(H) \otimes \overline{H}$, *and one has the following (non-canonically) split short exact sequence*

$$1 \longrightarrow \ker(H) \otimes \overline{H} \longrightarrow \Gamma \rightleftarrows \Gamma_s \longrightarrow 1.$$

(ii) *The relation between* $\hat{\Gamma}$ *and* Γ *is given by the following (non-canonically) split short exact sequence*

$$1 \longrightarrow H' \otimes \overline{H} \longrightarrow \hat{\Gamma} \rightleftarrows \Gamma \longrightarrow 1,$$

where $H' \otimes \overline{H}$ *is a subgroup of* $\ker(\hat{H}) \otimes \overline{H}$ *acting on* $\hat{H}$ *as indicated above.*

The proofs of the above theorems will be given in Section 4.2.

Remark 4.1.4. For the hyperbolic singularities the number $\mu_0 = \operatorname{rk}(\ker H)$ has the following values:

$$\mu_0 = \begin{cases} 1 & \text{for } T_{p,q,r},\ T^2_{p,q,r,s}, \\ n-1 & \text{for } T^n_{2,q,2,s}. \end{cases}$$

The group Γ_s is the Coxeter group corresponding to one of the Coxeter graphs of Figure 4.1.1. This follows from [Gabrielov$_2$] for the singularities $T_{p,q,r}$, from [Looijenga$_1$, II. 3.7] for the singularities $T^2_{p,q,r,s}$, and from Chapter 3.2 for the singularities $T^n_{p,q,r,s}$.

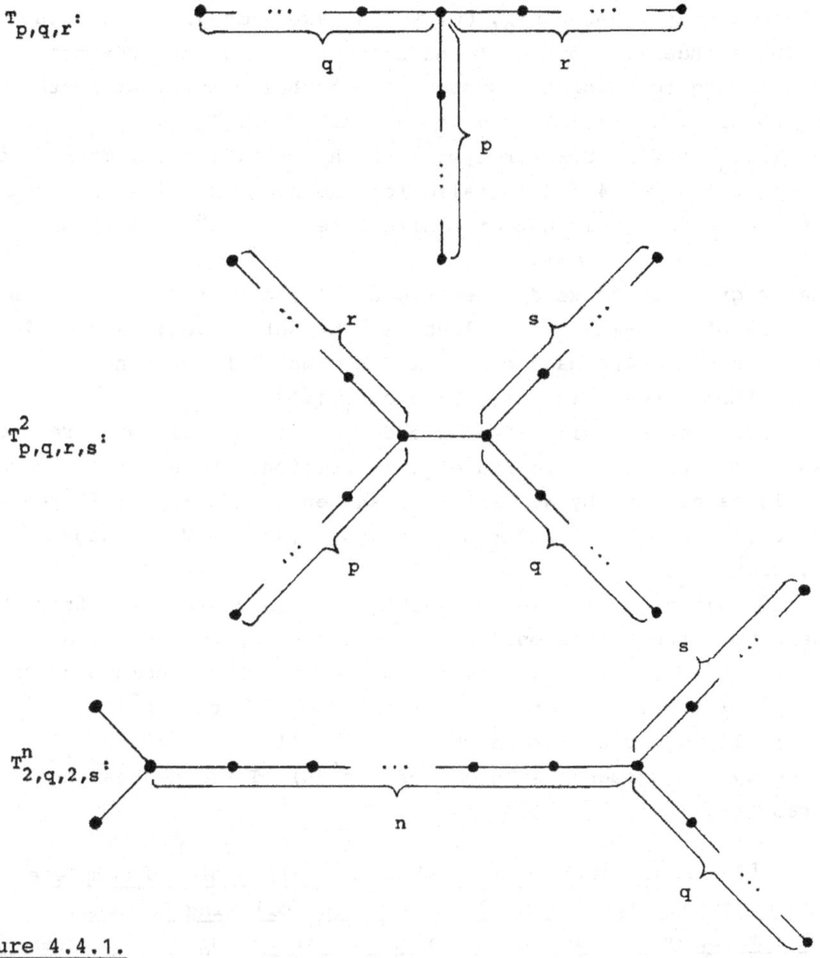

Figure 4.4.1.

The lattice $\overline{H}$ is the lattice corresponding to the Coxeter graph of Figure 4.1.1 associated with the singularity. We denote it by $Q_{p,q,r}$, $Q^2_{p,q,r,s}$, and $Q^2_{2,q,2,s}$ respectively. The discriminant of the lattice $Q_{p,q,r}$ is given in Chapter 3.4; the discriminants of the remaining lattices are given by the following formulas:

$$\operatorname{disc}(Q^2_{p,q,r,s}) = (-1)^{p+q+r+s}((p+r)(q+s) - pqrs) ,$$

$$\operatorname{disc}(Q^n_{2,q,2,s}) = \varepsilon^{q+s+n}\, 4(q+s-qs) .$$

According to [Bourbaki$_2$, Ch. V, §4, Exercice 12], the group Γ_s is of finite index in $O(\overline{H})$ if and only if Γ_s is the Coxeter group corresponding to a Coxeter system of hyperbolic type. Among the Coxeter graphs of Figure 4.1.1 only the graphs T_{237}, T_{245}, T_{334}, T^2_{2223}, and T^4_{2223} define Coxeter systems of hyperbolic type. Therefore assertion (i) of Theorem 4.1.2 is false for the singularities excluded in this theorem, and Γ_s is not of finite index in $O(\overline{H})$ in these cases.

A Coxeter graph as above defines a generalized root system $\overline{\Delta}$ in $\overline{H}$ in the sense of Kac-Moody-Lie algebras. The set of vanishing cycles Δ is in this case the preimage of $\overline{\Delta}$ in H under the mapping $H \longrightarrow \overline{H}$ (cf. [Looijenga$_3$, (7.23)], [Slodowy$_1$, 1.17]).

In the case when the dimension is odd one has the following results which we want to mention at the end of this section. These results were proved in full generality by W. Janssen [Janssen$_1$], after partial results by N. A'Campo [A'Campo$_1$], B. Wajnryb [Wajnryb$_1$] , and S.V. Chmutov [Chmutov$_1$], [Chmutov$_2$].

In the odd-dimensional case the lattice H is a skew-symmetric lattice. We denote the corresponding symplectic group by $Sp(H)$. As above, let $\tau : Sp(H) \longrightarrow \mathrm{Aut}\,(H^\#/j(H))$ be the canonical homomorphism. We define $Sp^\#(H) := \ker \tau$. Let G_H be the subgroup of $Sp^\#(H)$ consisting of all automorphisms which act trivially on $H^\#/2j(H)$. It contains a certain congruence subgroup of $Sp^\#(H)$. Then one has the following results.

<u>Theorem 4.1.5</u> [Janssen$_1$, (2.5) and (2.9)]. <u>For all isolated complete intersection singularities of odd dimension the following is true:</u>

(i) <u>The monodromy group</u> Γ <u>contains the group</u> G_H <u>and is therefore of finite index in</u> $Sp^\#(H)$.

(ii) <u>Except in the case of an ordinary double point, the following is true for the set of vanishing cycles</u> Δ : A vector $v \in H$ <u>is contained in</u> Δ <u>if and only if there exists a</u> $y \in H$ <u>and a</u> $\delta \in \Delta$ <u>with</u> $\langle v,y\rangle = 1$ <u>and</u> $v - \delta \in 2H$.

<u>Remark 4.1.6.</u> In the odd-dimensional case the monodromy group and the set of vanishing cycles do not only depend on the Milnor lattice, compare [Chmutov$_2$, 2.2].

4.2. Reduction to algebraic results

In this section we give the proofs of Theorems 4.1.1, 4.1.2, and 4.1.3. In particular we reduce the assertions of Theorem 4.1.2 to algebraic results about vanishing lattices, which will be proved in Chapter 5. The following proposition plays a central rôle for this reduction.

Proposition 4.2.1. Each isolated complete intersection singularity of dimension $n \geq 1$, which is not a simple hypersurface singularity and not of type $\tilde{E}_6$, $\tilde{E}_7$, $\tilde{E}_8$, $\tilde{D}_{n+3}$, $T_{p,q,r}$, $T^2_{p,q,r,s}$, or $T^n_{2,q,2,s}$, deforms to an exceptional unimodal hypersurface singularity, a singularity $J^{(n-1)}_{2n+5}$ (cf. Chapter 3.3), or a singularity T^n_{2233} $(n \geq 3)$ (cf. Chapter 3.1).

Proof. For the hypersurface singularities, the assertion can be deduced from the classification of Arnol'd, as was observed by D. Siersma (cf. [Brieskorn$_4$, p. 46]).

Now let $(X,0)$ be a singularity which is given by a map-germ $f : (\mathbb{C}^{n+k},0) \longrightarrow (\mathbb{C}^k,0)$ with $k \geq 2$, with an isolated singularity at 0, and with $df(0) = 0$. I am indebted to C.T.C. Wall for the following arguments.

We first consider the case when $k = 2$ and $n = 1$. If the 2-jet of f is regular, then we can apply the results of Chapter 3.1. If the corresponding non-degenerate pencil of quadrics has only eigenvalues of multiplicity less than or equal to 2, then the singularity is of type $\tilde{D}_4$ or $T^1_{2,q,2,s}$. Otherwise it deforms to a singularity with Segre symbol $\{3\}$, hence to a singularity of the J-series. These in their turn deform to the singularity J_7.

If the 2-jet of f is singular, then we have to consider two cases by [Wall$_4$, §4]. In the first case the corresponding pencil has rank 2, and the regular part has Segre symbol $\{1,1\}$. Then

$$(2xy, x^2 + y^2 + 2txz)$$

yields a deformation to the J-series. In the second case the corresponding pencil has rank 0. Then

$$(2xy, yz + tx^2)$$

yields a deformation to the J-series.

Now let still $k = 2$, but $n \geq 2$. If the 2-jet of f is singular or regular and one of the eigenvalues of the corresponding pencil of quadrics has multiplicity greater than or equal to 3, then f deforms to a map-germ

$$\tilde{f} \oplus (w_1^2, a_1 w_1^2) \oplus \ldots \oplus (w_{n-1}^2, a_{n-1} w_{n-1}^2)$$

with $\tilde{f} : (\mathbb{C}^3,0) \longrightarrow (\mathbb{C}^2,0)$, and with all a_i different and different from the eigenvalues of $j^2\tilde{f}$, where $\tilde{f}$ has the above properties. According to the case $n = 1$ considered above, $\tilde{f}$ deforms to the singularity J_7, hence f to the singularity $J_{2n+5}^{(n-1)}$. Otherwise, the pencil of quadrics corresponding to the 2-jet of f is regular, and all eigenvalues have multiplicity smaller than or equal to 2. If f is not of type $\tilde{D}_{n+3}$, $T^2_{p,q,r,s}$, or $T^n_{2,q,2,s}$, then $n \geq 3$ and at least two eigenvalues have exactly multiplicity 2. The simplest singularity with these properties is the singularity T^n_{2233}. Hence f deforms to this singularity.

Finally, we consider the case when $k \geq 3$, $n \geq 1$. We may assume that $k = 3$, since otherwise we can deform f in such a way that $k - 3$ component functions become linear. Therefore let $k = 3$. We show that f has a perturbation f_t such that the map-germ f_t, for $t \neq 0$, defines the same singularity as a map-germ $\tilde{f}_t : (\mathbb{C}^{n+2},0) \longrightarrow (\mathbb{C}^2,0)$ with 2-jet having an eigenvalue of multiplicity greater than or equal to 3.

Let $z = (z_0, z_1, \ldots, z_{n+2})$ be the coordinates of $\mathbb{C}^{n+3}$, and $z' = (z_1, \ldots, z_{n+2})$. Let $j^2 f = (q_1, q_2, q_3)$, and $q_i'(z_1, \ldots, z_{n+2}) = q_i (0, z_1, \ldots, z_{n+2})$ for $i = 1,2,3$. Denote the matrix of q_i' by Q_i'. Then the homogeneous equation

$$\det(\xi_1 Q_1' + \xi_2 Q_2' + \xi_3 Q_3') = 0$$

defines a curve $\Sigma \subset \mathbb{P}^2$ of degree $n + 2 \geq 3$. Let

$$a_1 \xi_1 + a_2 \xi_2 + a_3 \xi_3 = 0$$

be the equation of a line in $\mathbb{P}^2$ which meets the curve Σ in a point with intersection multiplicity greater than or equal to 3, e.g. an inflection tangent. Then we consider the perturbation

$$f_t(z) = (f_1(z) + t a_1 z_0,\ f_2(z) + t a_2 z_0,\ f_3(z) + t a_3 z_0).$$

This perturbation has the desired properties: Since $a_i \neq 0$ for at least one index i, we may assume $a_3 \neq 0$ without loss of generality. By a linear change of coordinates in the target we can achieve that

$$j^2 f_t(z) = (a_3 q_1(z) - a_1 q_3(z),\ a_3 q_2(z) - a_2 q_3(z),\ q_3(z) + t a_3 z_0).$$

If we solve the equation $q_3(z) + t a_3 z_0 = 0$ for z_0 and substitute the expression for z_0 in f_t, then we get

$$j^2 f_t(z) = (a_3 q_1'(z') - a_1 q_3'(z'),\ a_3 q_2'(z') - a_2 q_3'(z'),\ q_3(z) + t a_3 z_0).$$

From the choice of the a_i, $i = 1,2,3$, it follows that the pencil of quadrics determined by $a_3 q_1'(z') - a_1 q_3'(z')$ and $a_3 q_2'(z') - a_2 q_3'(z')$ has at least one eigenvalue of multiplicity greater than or equal to 3. This implies the assertion of the proposition for f and hence completes the proof of Proposition 4.2.1.

Proof of Theorem 4.1.1. The assertion of Theorem 4.1.1 is well-known for hypersurface singularities (cf. [Arnol'd$_1$]).

Therefore we consider isolated singularities of complete intersections which are not hypersurfaces. The singularities of type $\tilde{D}_{n+3}$ are parabolic according to Chapter 2.3. The singularities of type $T^2_{p,q,r,s}$ and $T^n_{2,q,2,s}$ are hyperbolic by Remark 4.1.4.

One has $\mu_{-\varepsilon} = 2$ for the singularities $J^{(n-1)}_{2n+5}$ according to Chapter 3.3, and for the singularities T^n_{2233} according to Chapter 2.5. By Proposition 4.2.1 all the other singularities deform to these singularities. Then [Looijenga$_3$, Proposition (7.13)] implies $\mu_{-\varepsilon} \geq 2$ for all the other singularities. This proves Theorem 4.1.1.

Proof of Theorem 4.1.2. We distinguish between four cases corresponding to the values of $\mu_{-\varepsilon}$ and μ_0.

Case (A) : $\mu_{-\varepsilon} = 0$, $\mu_0 = 0$. In this case the Milnor lattice is ε-definite, and (X,x) is a simple hypersurface singularity. The sets of vanishing cycles and the monodromy groups are finite in these cases. They are the root systems and Weyl groups respectively of type A_μ, D_μ, E_6, E_7, or E_8. One can easily check the assertions (i), (ii) and (iii) in these cases.

Case (B) : $\mu_{-\varepsilon} = 0$, $\mu_0 \neq 0$. In this case the Milnor lattice is parabolic. The corresponding singularities are the singularities of type

$\tilde{E}_8$, $\tilde{E}_7$, $\tilde{E}_6$, or $\tilde{D}_{n+3}$. For these singularities the assertions (i), (ii), and (iii) are also known (cf. [Looijenga$_3$, (7.23)(α)]), or they easily follow from Chapter 2.3.

Case (C) : $\mu_{-\varepsilon} = 1$. Concerning the hyperbolic singularities, we only have to check the assertions for the singularities T_{237}, T_{245}, T_{334}, T^2_{2223}, and T^4_{2223}. By Theorem 4.1.3 and Remark 4.1.4 the monodromy group Γ is the semi-direct product of the group $\ker(H) \otimes \bar{H}$ with the Coxeter group $W = \Gamma_s$ corresponding to the Coxeter graph associated with the singularity. It suffices to show that one has $W = O^*_\varepsilon(\bar{H})$ and $\bar{\Delta} = \{v \in \bar{H} \mid \langle v,v \rangle = 2\varepsilon\}$, where $\bar{\Delta}$ is the image of Δ in $\bar{H}$. We refer to Chapter 5.5 for a proof of these assertions.

Case (D) : $\mu_{-\varepsilon} \geq 2$. In this case we deduce the assertions of Theorem 4.1.2 from the algebraic results of Chapter 5. For this purpose we have to show that the vanishing lattices (H,Δ) and $(\hat{H},\hat{\Delta})$ of a singularity (X,x) with $\mu_{-\varepsilon} \geq 2$ are complete in the sense of Definition 5.3.1. Then assertion (i) follows from Theorems 5.3.2 and 5.3.3, assertion (ii) from (i) and Lemma 4.2.2, and assertion (iii) from Proposition 5.3.5.

By Proposition 4.2.1 and Theorem 4.1.1 every singularity (X,x) with $\mu_{-\varepsilon} \geq 2$ deforms to a singularity (Y,y), where (Y,y) is an exceptional unimodal hypersurface singularity or a singularity of type $J^{(n-1)}_{2n+5}$ or T^n_{2233}. According to [Looijenga$_3$, Proposition (7.13)] the vanishing lattice (H_X,Δ_X), respectively $(\hat{H}_X,\hat{\Delta}_X)$, of (X,x) contains the vanishing lattice (H_Y,Δ_Y), respectively $(\hat{H}_Y,\hat{\Delta}_Y)$, of (Y,y). By Remark 5.3.8 it suffices to show that the vanishing lattices (H_Y,Δ_Y) and $(\hat{H}_Y,\hat{\Delta}_Y)$ of (Y,y) are complete.

The exceptional unimodal hypersurface singularities have Dynkin diagrams of type $\tilde{\mathcal{O}}_{p,q,r}$, where $\tilde{\mathcal{O}}_{p,q,r}$ is the graph of Chapter 3.4. These Dynkin diagrams correspond to weakly distinguished and even strongly distinguished bases of vanishing cycles by [Gabrielov$_2$] and [Ebeling$_1$] respectively. These bases are special in the sense of Definition 5.4.1. The singularities $J^{(n-1)}_{2n+5}$ and T^n_{2233} have special weakly distinguished bases of thimbles and special weakly distinguished systems of generators of vanishing cycles by Remark 3.3.5 and Chapter 2.5 respectively. Moreover, the corresponding Milnor lattices contain sublattices of type $A_2 \perp U \perp U$. Therefore Theorem 5.4.2 implies that the assertion (ii) of Corollary 5.3.6 is valid. Then it follows from this corollary that the corresponding vanishing lattices (H_Y,Δ_Y) and $(\hat{H}_Y,\hat{\Delta}_Y)$ are complete. This proves Theorem 4.1.2.

Lemma 4.2.2. Let L be an even lattice, and L_1 a sublattice of L with $L_1 \subset \ker L$. Then for a given orthogonal splitting $L = L_1 \perp K$, there exists a corresponding split short exact sequence

$$1 \longrightarrow L_1 \otimes \bar{K} \longrightarrow O^*_\varepsilon(L) \rightleftarrows O^*_\varepsilon(K) \longrightarrow 1.$$

Proof. First the given splitting $L = L_1 \perp K$ leads to the following split short exact sequence

$$1 \longrightarrow L_1 \otimes K^\# \longrightarrow O(L) \rightleftarrows O(K) \times O(L_1) \longrightarrow 1.$$

Here $v \otimes \varphi \in L_1 \otimes K^\#$ acts on L_1 trivially, and on K by $(v \otimes \varphi)(y) = y + \varphi(y)v$ for $y \in K$. For $O^*_\varepsilon(L) \subset O(L)$ one obtains from this the above sequence (cf. also [Ebeling$_1$,§3]).

Proof of Theorem 4.1.3. For the non-hyperbolic singularities, the assertions (i) and (ii) follow from Theorem 4.1.2 (i) and Lemma 4.2.2.

For the hyperbolic singularities, the assertion (i) is known except for the singularities $T^n_{2,q,2,s}$, see [Gabrielov$_2$], [Looijenga$_3$, (7.23)]. The assertions (i) and (ii) are proved for the hyperbolic singularities with the aid of the given Dynkin diagrams. It follows from these Dynkin diagrams that the kernel $\ker \hat{H}$ of $\hat{H}$ is generated by differences of vanishing cycles (cf. Chapters 2.5 and 2.4). Let L be an even lattice with $L = L_1 \perp K$ and $L_1 \subset \ker L$. Then one has the following formula for $v \otimes w\ L_1 \otimes \bar{K}$ with $\langle w,w\rangle = 2\varepsilon$, $\varepsilon = \pm 1$:

$$(v \otimes w)(y) = (s_w \circ s_{w-\varepsilon v})(y) \quad \text{for } y \in L$$

$(v \otimes w)$ acts as the Eichler-Siegel transformation $\psi_{-v,w}$, cf. Chapter 5.1). Using these facts, one can deduce the assertions of Theorem 4.1.3 from the given Dynkin diagrams.

4.3. Global monodromy groups and Lefschetz pencils

From the descriptions of the local invariants in Section 4.1 one can also derive results for global invariants of projective complete intersections. Here we consider the case of even dimension; for the case of odd dimension see [Janssen$_1$],[Beauville$_1$].

For $d = (d_1,\dots,d_k)$ let $S_{n,d}$ be the projective space which parametrizes the complete intersections of k hypersurfaces of degrees $d_1,\dots,d_k$ in the complex projective space $\mathbb{P}^{n+k}$. The smooth complete intersections correspond to an open subset $U = U_{n,d}$ of this space.

Let $\pi : \mathcal{Y} \longrightarrow U$ be the corresponding family of smooth projective complete intersections, and denote by Y_u the fibre over a base point $u \in U$. The fundamental group $\pi_1(U,u)$ acts on the primitive cohomology group $P^n(Y_u,\mathbb{Z})$. The image $\Gamma_{n,d}$ of this representation $\rho : \pi_1(U,u) \longrightarrow \mathrm{Aut}(P^n(Y_u,\mathbb{Z}))$ is called the <u>global monodromy group</u> of the universal family of projective complete intersections of dimension n and multidegree $d = (d_1,\ldots,d_k)$. As in [Ebeling$_3$, § 6] for hypersurfaces, one can derive more generally for complete intersections from Theorem 4.1.2:

<u>Corollary 4.3.1. Let n be even, and $\varepsilon = (-1)^{n/2}$. Then the global monodromy group $\Gamma_{n,d}$ coincides with the subgroup $O^*_\varepsilon(P^n(Y_u,\mathbb{Z}))$ of the group of units of the non-degenerate lattice $P^n(Y_u,\mathbb{Z})$, hence is of finite index in $O(P^n(Y_u,\mathbb{Z}))$, and therefore arithmetic.</u>

The vanishing cycles and thimbles were originally introduced by Lefschetz in the context of Lefschetz pencils [Lefschetz$_1$]. We can apply our results to this situation, too. We briefly explain this situation. We refer to [Lamotke$_1$] for details.

A pencil of hyperplanes in $\mathbb{P}^N$ consists of all hyperplanes of $\mathbb{P}^N$ which contain a fixed $(N-2)$-dimensional projective subspace A called the <u>axis</u> of the pencil. Such a pencil corresponds to a line ℓ in the dual projective space $\check{\mathbb{P}}^N$, and we denote the pencil by $\{H_t\}_{t\in\ell}$. Let V be a non-singular subvariety of $\mathbb{P}^N$ of dimension $n+1$. We consider the intersections $V_t = V \cap H_t$ for $t \in \ell$. One calls $\{V_t\}_{t\in\ell}$ a <u>pencil of hyperplane sections</u> of V. The intersection $V \cap H$ for a hyperplane H is non-singular if and only if H is not tangent to V. The hyperplanes which are tangent to V correspond to the points of the dual variety $\check{V}$ of V in $\check{\mathbb{P}}^N$. The dual variety is in general a hypersurface with singularities. We now require that the line ℓ intersects the dual variety $\check{V}$ transversally and only in regular points. This condition is satisfied for a generic line ℓ. If this condition is satisfied, then we call the pencil $\{V_t\}_{t\in\ell}$ of hyperplane sections a <u>Lefschetz pencil</u> of hyperplane sections. Then the axis A of the corresponding pencil of hyperplanes intersects the variety V transversally. Let Y be the space which one obtains by blowing up V along the intersection $V \cap A$ of the axis of the pencil with V. Then Y is non-singular, and one gets a mapping $F : Y \longrightarrow \ell$ with the following properties: The fibres of F are the hyperplane sections V_t, and the critical values of F are the (finitely many) points of $\check{V} \cap \ell$. Every singular fibre has exactly one singular point, and this is non-degenerate, hence an ordinary double point.

Define $\ell^* := \ell - (\check{V} \cap \ell)$ and $Y^* = F^{-1}(\ell^*)$. Then $F : Y^* \longrightarrow \ell^*$ is a locally trivial fibre bundle.

Let $\mathbb{D}$ be a closed disc in ℓ which contains the critical values of F in its interior, and let t be a point on the boundary of $\mathbb{D}$. Set $Y_{\bar{\mathbb{D}}} := F^{-1}(\bar{\mathbb{D}})$ and $Y_t = F^{-1}(t)$. According to Chapter 1, one can define the set of vanishing cycles $\Delta \subset H_n(Y_t,\mathbb{Z})$, the set of thimbles $\hat{\Delta} \subset H_{n+1}(Y_{\bar{\mathbb{D}}},Y_t)$, and the corresponding groups $\Gamma_\Delta \subset \mathrm{Aut}(H_n(Y_t,\mathbb{Z}))$ and $\Gamma_{\hat{\Delta}} \subset \mathrm{Aut}(H_{n+1}(Y_{\bar{\mathbb{D}}},Y_t))$ respectively. The vanishing cycles $\delta \in \Delta$ generate the subgroup $P_n(Y_t,\mathbb{Z})$ of $H_n(Y_t,\mathbb{Z})$, the <u>primitive</u> or <u>vanishing</u> homology group. This group is Poincaré dual to the primitive cohomology group $P^n(Y_t,\mathbb{Z})$. The thimbles $\hat{\delta} \in \hat{\Delta}$ generate the relative homology group $H_{n+1}(Y_{\bar{\mathbb{D}}},Y_t)$, which has rank m, where m is the number of critical values of F. Any two vanishing cycles can be transformed into each other up to a sign by the group Γ_Δ [Lamotke$_1$, (7.3.5)]. Hence in the even-dimensional case, Δ is an orbit under Γ_Δ. In the odd-dimensional case this follows provided that there exist $\delta_1,\delta_2 \in \Delta$ with $\langle\delta_1,\delta_2\rangle = 1$ [Janssen$_1$, Lemma (2.1)]. This is also one of the conditions which have to be satisfied by a vanishing lattice (L,Δ) (see Chapter 1.3). Therefore, if there exist $\delta_1,\delta_2 \in \Delta$ with $\langle\delta_1,\delta_2\rangle = 1$ (except in the case when $\mathrm{rk}\, P_n(Y_t,\mathbb{Z}) = 1$), then it follows that $(P_n(Y_t,\mathbb{Z}),\Delta)$ and $(H_{n+1}(Y_{\bar{\mathbb{D}}},Y_t),\hat{\Delta})$ are vanishing lattices.

Now let V be a complete intersection, and let $\{V_t\}_{t\in\ell}$ be a Lefschetz pencil of hyperplane sections of V, such that V_t is a complete intersection of dimension n and multidegree $d = (d_1,\dots,d_k)$ in $\mathbb{P}^{n+k}$ for all $t \in \ell$. If V is sufficiently general, then a theorem of Zariski implies that the inclusion $\ell^* \subset U_{n,d}$ induces an epimorphism of the corresponding fundamental groups. Instead of considering the action of the fundamental group $\pi_1(U_{n,d},t)$ on the primitive cohomology group $P^n(Y_t,\mathbb{Z})$ as above, we can alternatively regard the action on the primitive homology group $P_n(Y_t,\mathbb{Z})$. The image of this representation in $O(P_n(Y_t,\mathbb{Z}))$ is also denoted by $\Gamma_{n,d}$. Then it follows that the monodromy group $\Gamma_{n,d}$ coincides with the group Γ_Δ.

There is the following relation to the local situation. Let Y_0 be a complete intersection of k hypersurfaces of degrees $d_1,\dots,d_k$ in $\mathbb{P}^{n+k}$ with an isolated singularity $x \in Y_0$. The Milnor fibre X_η of this singularity is the intersection of a smooth fibre Y_η close to Y_0 with a small ball centred at x in $\mathbb{P}^{n+k}$. Let $H_x = H_n(X_\eta,\mathbb{Z})$, and let $\Delta_x \subset H_x$ be the corresponding set of vanishing cycles. Transport from Y_η to Y_t yields a homomorphism $\iota : H_x \longrightarrow P_n(Y_t,\mathbb{Z})$ which preserves the intersection forms and maps Δ_x to Δ. If the intersection form on H_x is non-degenerate, then this homomorphism is injective.

Now let n be even. Using the above relation, one can then show that the pairs $(P_n(Y_t,\mathbb{Z}),\Delta)$ and $(H_{n+1}(Y_{\overline{D}},Y_t),\hat{\Delta})$ are complete vanishing lattices for $k = 1$, $d \geq 3$, and $n \geq 4$, and for $k \geq 2$ and $d \neq (2,2)$ [Beauville$_1$]. For this purpose one has to find suitable singularities (Y_0,x) with complete vanishing lattices (H_x,Δ_x), where the lattices H_x are non-degenerate. For example it is shown in [Beauville$_1$] that for given n,d, and k as above one can always find a complete intersection of dimension n and multidegree d in $\mathbb{P}^{n+k}$ which has an exceptional unimodal hypersurface singularity of type U_{12} in Arnol'd's notation. Hence it follows that the descriptions of Theorem 4.1.2 for the sets Δ and $\hat{\Delta}$ and the groups $\Gamma_\Delta = \Gamma_{n,d}$ and $\Gamma_{\hat{\Delta}}$ are also valid in this case, replacing H and $\hat{H}$ by $P_n(Y_t,\mathbb{Z})$ and $H_{n+1}(Y_{\overline{D}},Y_t)$ respectively. The remaining cases, when the above conditions on the numbers n,d, and k are not satisfied, are the quadrics, the cubic surfaces and the intersections of two quadrics. In these cases the lattices $P_n(Y_t,\mathbb{Z})$ are definite and correspond to the lattices A_1, E_6, and D_{n+3} respectively. It is easy to check that the descriptions of Theorem 4.1.2, except assertion (iii) for the quadrics, are also valid for these complete intersections.

Remark 4.3.2. The index in Corollary 4.3.1 can be specified. The lattice $L = P^n(Y_u,\mathbb{Z})$ is non-degenerate. It follows from [Nikulin$_1$, 1.14.2] that the homomorphism $\tau : O(L) \longrightarrow O(L^\#/L)$ is surjective for indefinite L. This can also be shown for the above definite lattices. For a reflection s_v corresponding to a vector v with $\langle v,v\rangle = -2\varepsilon$ one has $\sigma_\varepsilon(s_v) = -1$, but $s_v \in \ker \tau$. Therefore the index of $\Gamma_{n,d}$ in $O(L)$ is equal to the order of $O(L^\#/L)$ multiplied by 2, provided that L is indefinite.

Example 4.3.3. For $k = 1$, $n = 2$, and $d = 4$ (quartic surfaces in $\mathbb{P}^3$), $\Gamma_{n,d}$ has index 4 in $O(L)$.

5. MONODROMY GROUPS OF SYMMETRIC VANISHING LATTICES

5.1. Units of lattices

Let L be an even lattice, i.e. a free finitely generated $\mathbb{Z}$-module with a symmetric bilinear form $<,>$ with $<x,x> \in 2\mathbb{Z}$ for all $x \in L$. We denote the group of units (= isometries) of L by $O(L)$. Let $\varepsilon = \pm 1$ be fixed throughout this chapter. A vector $\delta \in L$ with $<\delta,\delta> = 2\varepsilon$ is called a minimal vector of square length 2ε. The reflection s_δ corresponding to such a vector δ is defined by

$$s_\delta(x) = x - \frac{2<x,\delta>}{<\delta,\delta>}\delta \qquad \text{for} \quad x \in L.$$

Let $\Delta \subset L$ be a subset of L with $<\delta,\delta> = 2\varepsilon$ for all $\delta \in \Delta$. We denote the sublattice generated by Δ by $\mathbb{Z}\cdot\Delta$, and the subgroup of $O(L)$ generated by all reflections s_δ for $\delta \in \Delta$ by Γ_Δ. We define $R_\varepsilon(L)$ to be the subgroup of $O(L)$ generated by <u>all</u> reflections s_v corresponding to minimal vectors v of square length 2ε.

We define a subgroup $O^*_\varepsilon(L)$ of $O(L)$ as follows. We denote the kernel (radical) of L by ker L and define $\bar{L} = L/\ker L$. The lattice $\bar{L}$ is the non-degenerate lattice corresponding to L. Moreover, we set $\bar{L}_{\mathbb{R}} = \bar{L} \otimes \mathbb{R} = \mathbb{R}.\bar{L}$. We define a homomorphism

$$\sigma_\varepsilon : O(L) \longrightarrow \{+1,-1\}$$

(real spinor norm) as follows. Let $g \in O(L)$ and let $\bar{g}$ be the induced element in $O(\bar{L}) \subset O(\bar{L}_{\mathbb{R}})$. Then one can write $\bar{g}$ in $O(\bar{L}_{\mathbb{R}})$ as a product of reflections $\bar{g} = s_{v_1} \circ \ldots \circ s_{v_r}$. We define

$$\sigma_\varepsilon(g) := \begin{cases} +1 & \text{if} \quad \varepsilon<v_i,v_i> < 0 \quad \text{for an even number of indices,} \\ -1 & \text{otherwise.} \end{cases}$$

We denote the dual lattice $\mathrm{Hom}(L,\mathbb{Z})$ by $L^\#$. Let $j : L \longrightarrow L^\#$ be the natural homomorphism. Then one has a canonical homomorphism

$$\tau : O(L) \longrightarrow \mathrm{Aut}(L^\#/jL).$$

<u>Definition 5.1.1.</u> $O^*_\varepsilon(L) := \ker \sigma_\varepsilon \cap \ker \tau$.

Now let $f \in L$ be an isotropic vector (i.e. $<f,f> = 0$), and let $w \in L$ be a vector orthogonal to f. Then the <u>Eichler-Siegel trans-</u>

<u>formation</u> $\psi_{f,w} \in O(L)$ is defined by

$$\psi_{f,w}(x) = x + \langle x,f\rangle w - \langle x,w\rangle f - \frac{1}{2}\langle w,w\rangle\langle x,f\rangle f$$

for $x \in L$ [Eichler$_1$]. Let U be a unimodular hyperbolic plane with a basis $\{f_1,f_2\}$ of isotropic vectors with $\langle f_1,f_2\rangle = 1$. We assume that L is the orthogonal direct sum of U and an even lattice M, hence $L = M \perp U$.

<u>Definition 5.1.2.</u> Let $\Psi_U(L)$ be the subgroup of $O(L)$ generated by the transformations $\psi_{f_1,w}$ and $\psi_{f_2,w}$ for arbitrary $w \in M$.

We list some properties of these transformations (cf. [Eichler$_1$], [Ebeling$_3$]):

(a) Every element $\varphi \in \Psi_U(L)$ has determinant 1 and spinor norm $\sigma_+(\varphi) = \sigma_-(\varphi) = 1$.

(b) For $w,w' \in M$ and $i = 1,2$ one has

$$\psi_{f_i,w} \circ \psi_{f_i,w'} = \psi_{f_i,w+w'},$$

hence $w \longmapsto \psi_{f_i,w}$ defines a homomorphism $M \longrightarrow \Psi_U(L)$.

(c) The following formulas are valid for $w \in M$ with $\langle w,w\rangle = 2\varepsilon$:

(c1) $\psi_{f_i,w} = s_w \circ s_{w+\varepsilon f_i}$,

(c2) $s_w = \psi_{f_2,w} \circ \psi_{f_1,\varepsilon w} \circ \psi_{f_2,w} \circ s_{f_1+\varepsilon f_2}$.

(d) $O(U)$ normalizes $\Psi_U(L)$.

(e) Let $L = U' \perp U$ for another unimodular hyperbolic plane U'. Then for each vector $x \in U' \perp U$ there exists a $\varphi \in \Psi_U(U' \perp U)$ with $\varphi(x) = \alpha f_1 + \beta f_2$, where $\alpha | \beta$ (possibly $\beta = 0$).

Property (e) directly implies the following lemma, which will play an important rôle.

<u>Lemma 5.1.3.</u> <u>Let</u> $L = L'' \perp U' \perp U$. <u>Then for each vector</u> $v \in L$ <u>there exists a</u> $\varphi \in \Psi_U(U' \perp U)$ <u>with</u> $\varphi(v) \in L'' \perp U'$.

Here we regard $\Psi_U(U' \perp U)$ as a subgroup of $\Psi_U(L)$ in a natural way.

Let Δ be the set of minimal vectors of square length 2ε. Then one has the following inclusions between the various groups:

$$\Gamma_\Delta \subset R_\varepsilon(L) \subset O^*_\varepsilon(L) \subset O(L),$$

$$\Psi_U(L) \subset O^*_\pm(L).$$

Property (b) and formula (c1) imply that one has in addition

$$\Psi_U(L) \subset R_\varepsilon(L),$$

if M is generated by minimal vectors of square length 2ε.

5.2. Symmetric vanishing lattices

We consider the symmetric analogue corresponding to a skew-symmetric vanishing lattice. The notion of a skew-symmetric vanishing lattice was introduced in [Janssen$_1$]. Let $\varepsilon \in \{+1,-1\}$ be fixed.

Definition 5.2.1. Let L be an even lattice, and let $\Delta \subset L$ be a subset of L. A pair (L,Δ) is called a (symmetric) vanishing lattice, if the following conditions are satisfied:

(i) The set Δ consists of minimal vectors of square length 2ε.

(ii) The set Δ generates L.

(iii) The set Δ is a Γ_Δ-orbit.

(iv) Unless $rk(L) = 1$, there exist $\delta_1, \delta_2 \in \Delta$ with $\langle\delta_1, \delta_2\rangle = 1$.

The group Γ_Δ is called the monodromy group of the vanishing lattice (L,Δ).

Example 5.2.2. The symmetric vanishing lattices (L,Δ) with ε-definite lattice L are exactly the pairs (L,Δ) where Δ is an irreducible homogeneous root system, hence of type A_m, D_m, E_6, E_7, or E_8, and L is the corresponding root lattice. Here, in the case when $\varepsilon = -1$, the usual bilinear form of [Bourbaki$_2$] has to be multiplied by $\varepsilon = -1$. We denote the corresponding lattices also by the symbols A_m, D_m, etc., where we make the convention that the usual bilinear form has to be multiplied by ε. The monodromy groups Γ_Δ are just the corresponding Weyl groups. This characterization of the definite vanishing lattices follows from [Bourbaki$_2$, Ch. VI, §4, especially n°4]. The condition (iii) for (L,Δ) just implies the irreducibility of the root system Δ. Hence the ε-definite vanishing lattices are exactly the Milnor lattices of the simple hypersurface singularities in dimension

$$n \equiv \begin{cases} 0 \pmod 4 & \text{for} \quad \varepsilon = +1 , \\ 2 \pmod 4 & \text{for} \quad \varepsilon = -1 . \end{cases}$$

Remark 5.2.3. If the set Δ or the monodromy group Γ_Δ of a vanishing lattice (L,Δ) is finite, then the lattice L is definite. I am indebted to K. Saito for the following simple proof of this fact (see also [Saito$_1$, (1.3) Note 2]). If Δ is finite, then also Γ_Δ is finite, since Γ_Δ can be regarded as a subgroup of the permutation group of Δ. Therefore let Γ_Δ be finite, and let $b : L \times L \longrightarrow \mathbb{Z}$ be any (positive or negative) definite symmetric bilinear form on L. Define

$$b'(x,y) = \sum_{\gamma \in \Gamma_\Delta} b(\gamma x, \gamma y), \quad \text{for} \quad x,y \in L.$$

Then b' is a definite symmetric bilinear form which is invariant under Γ_Δ. According to [Bourbaki$_2$, Ch. V, §2, Proposition 1], the bilinear form b' coincides with $\langle , \rangle$ up to a sign.

Example 5.2.4. Let $\varepsilon = \pm 1$ be fixed, and let Q be an even lattice which has a basis $B = \{e_1, \ldots, e_r\}$ with $\langle e_i, e_i \rangle = 2\varepsilon$ and

$$\langle e_i, e_j \rangle = -\varepsilon \quad \text{or} \quad 0 \quad \text{for} \quad 1 \leq i \neq j \leq r.$$

Here we assume that the Dynkin diagram corresponding to B is a connected graph. Let $S = \{s_{e_i} \mid i = 1, \ldots, r\}$, and let W be the subgroup of $O(Q)$ generated by the set S of reflections. Then (W,S) is an irreducible Coxeter system (cf. [Bourbaki$_2$, Ch. V, §4]). We define $R := W \cdot B = \{w(e_i) \mid w \in W, e_i \in B\}$. Then (Q,R) is a vanishing lattice, and the corresponding monodromy group Γ_R coincides with the Coxeter group W. If Q is ε-definite, then we are in the case of Example 5.2.2 and the corresponding vanishing lattices (Q,R) are just the vanishing lattices corresponding to the classical irreducible homogeneous root systems R. If Q is ε-semi-definite, then the corresponding vanishing lattices (Q,R) are just the vanishing lattices corresponding to the associated affine root systems R [Bourbaki$_2$, Ch. VI, § 4.3]. If Q is indefinite, then the group W is in general not of finite index in $O(Q)$. For suppose that Q is non-degenerate and indefinite. Then according to [Bourbaki$_2$, Ch. V, §4, Exercice 12], the index of W in $O(Q)$ is only finite if (W,S) is a Coxeter system of hyperbolic type. This means that the lattice Q is hyperbolic and that each proper subset of B generates a semi-definite sublattice of Q. We shall consider such vanishing lattices in Section 5.5.

5.3. Complete vanishing lattices

We now introduce additional conditions on a vanishing lattice.

Again let $\varepsilon = \pm 1$ be fixed as above. We first consider the lattice $A_2 \perp U' \perp U$. Let $\{\omega_1, \omega_2\}$ be a basis of A_2 consisting of minimal vectors of square length 2ε with $\langle\omega_1, \omega_2\rangle = -\varepsilon$. Let $\{f_1, f_2\}$ and $\{f_1', f_2'\}$ be bases of U and U' respectively consisting of isotropic vectors with $\langle f_1, f_2\rangle = \langle f_1', f_2'\rangle = 1$. We set

$$\Omega := \{\omega_1, \omega_2, \omega_1 - f_1, f_1 + \varepsilon f_2, \omega_1 - f_1', f_1' + \varepsilon f_2'\} .$$

The set Ω forms a basis of $A_2 \perp U' \perp U$ consisting of minimal vectors of square length 2ε .

<u>Definition 5.3.1.</u> A vanishing lattice (L, Δ) is called <u>complete</u>, if the following conditions are satisfied:

(i) The lattice L contains a sublattice $A_2 \perp U' \perp U$.

(ii) The set Δ contains the subset Ω of this sublattice defined above.

We can now formulate the main result of this chapter.

<u>Theorem 5.3.2.</u> <u>Let (L, Δ) be a complete vanishing lattice. Then</u>

$$\Gamma_\Delta = R_\varepsilon(L) .$$

The proof of this theorem will be given in Section 5.4.

M. Kneser has proved the following theorem.

<u>Theorem 5.3.3 (M. Kneser).</u> <u>Let L be an even lattice, and assume that the following conditions on the corresponding non-degenerate lattice $\bar{L}$ are satisfied:</u>

(i) <u>The Witt index of $\bar{L}_{\mathbb{R}}$ is at least 2.</u>

(ii) <u>The lattice $\bar{L}$ contains a sublattice L_1 of rank $\operatorname{rk}(L_1) \geq 5$ whose discriminant $\operatorname{disc}(L_1)$ is not divisible by 3, as well as a sublattice L_2 with $\operatorname{rk}(L_2) \geq 6$ and $2 \nmid \operatorname{disc}(L_2)$.</u>

<u>Then</u>

$$R_{\varepsilon}(L) = O^*_{\varepsilon}(L).$$

This theorem follows from [Kneser$_1$, Satz 4], using the arguments in [Ebeling$_2$, Proof of Theorem 3.1,c)].

The assumptions of Theorem 5.3.3 are in particular satisfied for lattices L which contain a sublattice $A_2 \perp U' \perp U$, hence also for complete vanishing lattices. Combining Theorem 5.3.2 and Theorem 5.3.3, one therefore gets:

Theorem 5.3.4. Let (L,Δ) be a complete vanishing lattice. Then

$$\Gamma_{\Delta} = O^*_{\varepsilon}(L) .$$

Using the following proposition, this theorem implies that the set Δ of a complete vanishing lattice (L,Δ) is already determined by L.

Proposition 5.3.5. Let (L,Δ) be a vanishing lattice which satisfies the following conditions:

(i) There exists an orthogonal splitting $L = L'' \perp U' \perp U$.

(ii) One has the inclusion $\Psi_U(L) \subset \Gamma_{\Delta}$. (This condition is in particular satisfied, if $\Gamma_{\Delta} = O^*_{\varepsilon}(L)$.)

Then

$$\Delta = \{v \in L \mid \langle v,v\rangle = 2\varepsilon \quad \text{and} \quad \langle v,L\rangle = \mathbb{Z}\},$$

where $\langle v,L\rangle = \mathbb{Z}$ means that there exists a $y \in L$ with $\langle v,y\rangle = 1$.

Proof. This proposition follows from a more general result due to E. Looijenga (cf. [Brieskorn$_4$, 4.2]).

For the convenience of the reader we give a direct proof following the same lines as the proof of [Looijenga-Peters, Theorem (2.4)] (cf. also [Pjateckiĭ-Šapiro-Šafarevič, Appendix to § 6]). By the definition of a vanishing lattice it is clear that Δ is contained in the set on the right-hand side. Therefore let v be an element of the set on the right-hand side. We write $v = v' + v''$ with $v' \in U' \perp U$ and $v'' \in L''$. Let $\{f_1,f_2\}$ be a basis of U as above. According to property (e) of Section 5.1, there exists a $\varphi \in \Psi_U(U' \perp U)$ such that $\tilde{v}' = \varphi(v')$ satisfies $\langle \tilde{v}',f_2\rangle \mid \langle \tilde{v}',f_1\rangle$. But since $\varphi(v'') = v''$, it follows that $\tilde{v} = \varphi(v)$ also satisfies $\langle \tilde{v},f_2\rangle \mid \langle \tilde{v},f_1\rangle$. We can therefore assume that this is already true for v. Since $\langle v,L\rangle = \mathbb{Z}$, there exists a $y \in L$

with $\langle v,y\rangle = -1$. We write $y = y_1 + y_2$, where $y_1 \in L'' \perp U'$ and $y_2 \in U$. Then

$$\psi_{f_2,y_1}(v) = \alpha f_1 + \beta f_2 + v_1$$

with $v_1 \in L'' \perp U'$. But the greatest common divisor of α and β is equal to 1, since $\alpha = \langle v,f_2\rangle$ and

$$\beta = \langle v,f_1\rangle - \langle v,y_1\rangle - \frac{1}{2}\langle y_1,y_1\rangle\langle v,f_2\rangle$$

$$\equiv -\langle v,y_1\rangle \bmod \alpha \equiv -\langle v,y\rangle \bmod \alpha \equiv 1 \bmod \alpha .$$

According to property (e) of Section 5.1, after another application of an element of $\Psi_U(U' \perp U)$ we can assume that $\alpha = 1$. Then $\psi_{f_2,-v_1}$ maps this vector to a vector of the form $f_1 + \tilde{\varepsilon} f_2$. But

$$\langle f_1 + \tilde{\varepsilon} f_2,\ f_1 + \tilde{\varepsilon} f_2\rangle = \langle v,v\rangle = 2\varepsilon$$

implies that $\tilde{\varepsilon} = \varepsilon$. So each minimal vector v of square length 2ε with $\langle v,L\rangle = \mathbb{Z}$ can be mapped to the vector $f_1 + \varepsilon f_2$ by an element of $\Psi_U(L) \subset \Gamma_\Delta$. This proves the proposition.

Corollary 5.3.6. Let (L,Δ) be a vanishing lattice, and assume that L contains a sublattice $A_2 \perp U' \perp U$. Then the following statements are equivalent:

(i) The vanishing lattice (L,Δ) is complete.

(ii) The inclusion $\Psi_U(L) \subset \Gamma_\Delta$ is valid.

(iii) The set Δ contains all minimal vectors of square length 2ε of $A_2 \perp U' \perp U$.

Proof. The implication (i) ⇒ (ii) follows from Theorem 5.3.4.

The implication (ii) ⇒ (iii) follows from Proposition 5.3.5, if one notes that for a minimal vector $v \in M = A_2 \perp U' \perp U$ one has $\langle v,M\rangle = \mathbb{Z}$.

The implication (iii) ⇒ (i) is trivial.

Remark 5.3.7. Condition (i) of Definition 5.3.1, namely the existence of a sublattice $A_2 \perp U' \perp U$, is sufficient for Theorem 5.3.3, but not for Theorem 5.3.2, as is shown by the following example. We consider the non-degenerate lattice $L := A_2 \perp A_2 \perp U' \perp U$. Then there exists a basis B of L consisting of minimal vectors of square length 2ε such that the matrix of the bilinear form with respect to this basis is given by the Dynkin diagram of Figure 5.3.1 (note the convention of Chapter 1.5). Let (L,Δ) be the vanishing lattice which is defined by B according to Example 5.2.4. Since L is non-degenerate, indefinite, and non-hyperbolic, the group Γ_Δ is not of finite index in $O(L)$ according to Example 5.2.4. According to Theorem 5.3.4, the vanishing lattice (L,Δ) is therefore not complete.

Figure 5.3.1

Remark 5.3.8. Let (L,Δ_L) and (K,Δ_K) be vanishing lattices. We say that (L,Δ_L) contains the vanishing lattice (K,Δ_K) if K is a primitive sublattice of L and $\Delta_K \subset \Delta_L$. Let $M = A_2 \perp U' \perp U$, and let $\Omega \subset M$ be the above basis of M. We set $\overline{\Omega} = \Gamma_\Omega \cdot \Omega$. Then one can easily show that $(M,\overline{\Omega})$ is a vanishing lattice (cf. Example 5.4.3). Then Definition 5.3.1 can also be formulated as follows: A vanishing lattice (L,Δ) is complete if and only if it contains the vanishing lattice $(M,\overline{\Omega})$. By Corollary 5.3.6 the set $\overline{\Omega}$ consists of all minimal vectors of M of square length 2ε. Note that a vanishing lattice (L,Δ_L) which contains a complete vanishing lattice is itself complete.

5.4. Special subsets

In this section we shall prove Theorem 5.3.2. For this purpose we have to introduce the notion of a special subset. Let L be an even lattice and Λ be a subset of L. We define an equivalence relation of Λ, denoted by $\sim_\Lambda$, as follows. For $\lambda,\lambda' \in \Lambda$ the element λ is equivalent to λ', $\lambda \sim_\Lambda \lambda'$, if and only if there exists a sequence $\lambda = \lambda_0,\lambda_1,\ldots,\lambda_k = \lambda'$ with $\langle\lambda_{i-1},\lambda_i\rangle = -\varepsilon$ and $\lambda_i \in \Lambda$ for $i = 1,\ldots,k$, or if $\lambda = \lambda'$.

Definition 5.4.1. A subset $\Lambda \subset L$ is called special, if the following conditions are satisfied:

(i) The set Λ consists of minimal vectors of square length 2ε.

(ii) There exist $\lambda_1,\lambda_2,\lambda_3 \in \Lambda$ with $\langle\lambda_1,\lambda_2\rangle = -\varepsilon$, and with $\langle\lambda_1,\lambda\rangle = 0$ and $\langle\lambda_2,\lambda\rangle = \langle\lambda_3,\lambda\rangle$ for all $\lambda \in \Lambda$ with $\lambda \neq \lambda_1,\lambda_2$.

(iii) Let $\Lambda' = \Lambda - \{\lambda_1,\lambda_2\}$. Then $\lambda \sim_{\Lambda'} \lambda_3$ for all $\lambda \in \Lambda'$.

Let Λ be a special subset of L, and let $\Lambda' = \Lambda - \{\lambda_1,\lambda_2\}$ as above. We set

$$
\begin{aligned}
f_1 &= -\lambda_2 + \lambda_3, \\
f_2 &= \varepsilon\lambda_1 - \lambda_2 + \lambda_3, \\
U &= \mathbb{Z}.f_1 + \mathbb{Z}.f_2 .
\end{aligned}
$$

Then U is a unimodular hyperbolic plane, and

$$\mathbb{Z}.\Lambda = U \perp \mathbb{Z}.\Lambda' .$$

Theorem 5.4.2. *Let $\Lambda \subset L$ be a special subset. Then*

$$\Psi_U(\mathbb{Z}.\Lambda) \subset \Gamma_\Delta .$$

Moreover, if another unimodular hyperbolic plane U' is contained in $\mathbb{Z}.\Lambda'$, then

$$R_\varepsilon(\mathbb{Z}.\Lambda) = \Gamma_\Delta .$$

The proof of Theorem 5.4.2. is the same as the proof of [Ebeling$_2$, Theorem 3] applied to the lattice $K = \mathbb{Z}.\Lambda$, except that we do not have a basis B' of $K' = \mathbb{Z}.\Lambda'$, but only a system of generators Λ'. But the linear independence of the elements of B' is not used.

Example 5.4.3. Let Ω be the subset of the lattice $M = A_2 \perp U' \perp U$ defined in Section 5.3. Then Ω is special: Let $\lambda_1 = f_1 + \varepsilon f_2$, $\lambda_2 = \omega_1 - f_1$, $\lambda_3 = \omega_1$. One easily checks that all elements of Ω are equivalent with respect to $\sim_\Omega$. For $\omega \in \Omega' = \Omega - \{\lambda_1,\lambda_2\}$ one even has $\omega \sim_{\Omega'} \lambda_3$. Moreover $U' \subset \mathbb{Z}.\Omega'$. Theorem 5.4.2 therefore implies

$$\Psi_U(U' \perp U) \subset \Psi_U(\mathbb{Z}.\Omega) \subset \Gamma_\Omega .$$

Now let (L,Δ) be a complete vanishing lattice. The idea of the proof of Theorem 5.3.2 is to extend the subset $\Omega \subset \Delta$ to a special system of generators $\Lambda \subset \Delta$ of L and to apply Theorem 5.4.2 subsequently. The following lemmas serve for this purpose.

Lemma 5.4.4. Let (L,Δ) be a complete vanishing lattice. Then the set Δ is an equivalence class with respect to $\sim_\Delta$.

Proof. Each equivalence class $\tilde{\Delta}$ is a $\Gamma_{\tilde{\Delta}}$-orbit. For $\langle\delta,\delta'\rangle=-\varepsilon$ implies that $s_{\delta'} s_\delta(\delta')=\delta$. On the other hand $\Gamma_{\tilde{\Delta}}(\tilde{\Delta}) \subset \tilde{\Delta}$. To prove this, first note that if $\delta \in \tilde{\Delta}$ then also $s_\delta(\delta) = -\delta$ lies in $\tilde{\Delta}$. For there is a $\gamma \in \Gamma_\Delta$ with $\gamma(\omega_2) = \delta$, and

$$\{\gamma(\omega_2),\ \gamma(\omega_1 - f_1),\ \gamma(f_1 + \varepsilon f_2)\}$$

is the basis of a root system of type A_3 which is contained in $\tilde{\Delta}$. Now let $\delta,\delta' \in \tilde{\Delta}$. Then one can easily show by induction on the minimal length of a sequence $\delta = \delta_0, \delta_1, \ldots, \delta_k = \delta'$ with $\langle\delta_{i-1},\delta_i\rangle = 1$ and $\delta_i \in \tilde{\Delta}$, for $i = 1,\ldots,k$, that $s_\delta(\delta') \in \tilde{\Delta}$.

Since Γ_Δ preserves the symmetric bilinear form, Γ_Δ permutes equivalence classes. Now let $\tilde{\Delta}$ be an equivalence class, and $\delta \in \Delta$. Then it suffices to show that there exists a $\tilde{\delta} \in \tilde{\Delta}$ such that $s_\delta(\tilde{\delta}) = \tilde{\delta}$, hence $\langle\delta,\tilde{\delta}\rangle = 0$. For this implies that Γ_Δ leaves the equivalence class $\tilde{\Delta}$ invariant, hence $\tilde{\Delta} = \Gamma_\Delta . \tilde{\Delta} = \Delta$.

Let $\delta_1 \in \tilde{\Delta}$. Then there is a $\gamma \in \Gamma_\Delta$ with $\gamma(\omega_1) = \delta_1$. This implies that $\gamma(\Omega) \subset \tilde{\Delta}$. We may assume without loss of generality that $\delta_1 = \omega_1$, since we can replace $U' \perp U$ by $\gamma^{-1}(U' \perp U)$ and ω_1, ω_2 by $\gamma^{-1}(\omega_1)$, $\gamma^{-1}(\omega_2)$. Then $\Omega \subset \tilde{\Delta}$. Since the lattice $U' \perp U \subset \mathbf{Z}.\Omega$ is unimodular, we can write $L = L'' \perp U' \perp U$. By Lemma 5.1.3 there exists a $\varphi \in \Psi_U(U' \perp U) \subset \Gamma_\Omega$ with $\varphi(\delta) \in L'' \perp U$. Then

$$\langle\delta,\varphi^{-1}(f_1 + \varepsilon f_2)\rangle = \langle\varphi(\delta),\ f_1 + \varepsilon f_2\rangle = 0,$$

and $f_1 + \varepsilon f_2 \in \tilde{\Delta}$, $\varphi \in \Gamma_{\tilde{\Delta}}$. Since each equivalence class $\tilde{\Delta}$ is a $\Gamma_{\tilde{\Delta}}$-orbit, the lemma is proven.

Lemma 5.4.5. Let (L,Δ) be a complete vanishing lattice. Define

$$\Delta_0 := \{\delta \in \Delta \mid \langle\omega_1,\delta\rangle = -\varepsilon \ \underline{\text{or}}\ \delta = \omega_1\}.$$

<u>Then</u> $\Gamma_{\Delta_0} = \Gamma_\Delta$ <u>and</u> $L = \mathbb{Z}.\Delta_0$.

<u>Proof.</u> The proof is the same as that of Lemma (2.7) in [Janssen_1]. Let $\delta \in \Delta$. Let $\ell(\delta)$ denote the minimal length of a sequence $\omega_1 = \delta_0, \delta_1, \ldots, \delta_k = \delta$ with $\delta_i \in \Delta$ and $<\delta_{i-1}, \delta_i> = -\varepsilon$ for $i = 1,\ldots,k$, which exists according to Lemma 5.4.4. We prove by induction on $\ell(\delta)$ that $\delta \in \Gamma_{\Delta_0}.\omega_1$. If $\ell(\delta) = 0$ then $\delta = \omega_1$. Now let $k = \ell(\delta) > 0$, and let a sequence as above be given. By the induction hypothesis there exists a $\gamma \in \Gamma_{\Delta_0}$ with $\gamma(\delta_{k-1}) = \omega_1$. But

$$<\omega_1, \gamma(\delta_k)> = <\gamma(\delta_{k-1}), \gamma(\delta_k)> = <\delta_{k-1}, \delta_k> = -\varepsilon \;,$$

hence $s_{\gamma(\delta_k)} \in \Gamma_{\Delta_0}$. This implies

$$s_{\delta_k} = \gamma^{-1} s_{\gamma(\delta_k)} \gamma \in \Gamma_{\Delta_0} \;.$$

Therefore

$$\delta_k = s_{\delta_{k-1}} s_{\delta_k}(\delta_{k-1}) \in \Gamma_{\Delta_0}.\,\omega_1 \;.$$

But $\Delta = \Gamma_{\Delta_0}.\Delta_0$ implies that $\Gamma_{\Delta_0} = \Gamma_\Delta$ and $L = \mathbb{Z}.\Delta = \mathbb{Z}.(\Gamma_{\Delta_0}.\Delta_0) \subset \mathbb{Z}.\Delta_0$. This proves Lemma 5.4.5.

<u>Proposition 5.4.6.</u> <u>Let</u> (L,Δ) <u>be a complete vanishing lattice. Define</u>

$$\Lambda_0 := \{\delta \in \Delta \mid <f_1,\delta> = <f_2,\delta> = 0\},$$
$$\Lambda' := \{\delta \in \Lambda_0 \mid \delta \sim_{\Lambda_0} \omega_1\} \;,$$
$$\Lambda \;:= \Lambda' \cup \{\omega_1 - f_1, f_1 + \varepsilon f_2\} \;.$$

<u>Then</u> Λ <u>is a special subset of</u> L <u>with</u> $\Omega \subset \Lambda \subset \Delta$, $U' \subset \mathbb{Z}.\Delta'$, <u>and</u> $L = \mathbb{Z}.\Lambda$.

<u>Proof.</u> It is clear that Λ is special and that $\Omega \subset \Lambda$. Since U' is contained in $\mathbb{Z}.\Omega$, it is also contained in $\mathbb{Z}.\Lambda'$. Thus we only have to show that $L = \mathbb{Z}.\Lambda$. By the previous lemma it suffices to show that Δ_0 is contained in $\mathbb{Z}.\Lambda$.

Let $\delta \in \Delta_0$, $\delta \neq \omega_1$. We have $L = L'' \perp U' \perp U$. By Lemma 5.1.3 there exists a $\varphi \in \Psi_U(U' \perp U)$ with $\varphi(\delta) \in L'' \perp U'$ and

$$<\varphi(\delta),\omega_1> = <\varphi(\delta),\varphi(\omega_1)> = <\delta,\omega_1> = -\varepsilon ,$$

since $\omega_1 \in L''$ and $\delta \in \Delta_0$. Thus $\gamma = \varphi(\delta)$ is contained in Λ'. But φ is an element of $\Gamma_\Omega \subset \Gamma_\Delta$. Hence

$$\delta = \varphi^{-1}(\gamma) \in \Gamma_\Lambda(\Lambda') \subset \mathbb{Z}.\Lambda .$$

This proves the proposition.

Theorem 5.3.2 now follows from Proposition 5.4.6 and Theorem 5.4.2.

5.5. Hyperbolic Coxeter systems

In this section we consider some non-complete vanishing lattices which are defined by hyperbolic Coxeter systems according to Example 5.2.4.

In Example 5.2.4 we defined vanishing lattices corresponding to certain Coxeter systems (W,S). Let (Q,R) be such a vanishing lattice. We already observed in Example 5.2.4 that if the lattice Q is non-degenerate and indefinite and the corresponding Coxeter system (W,S) is not of hyperbolic type, then the group $\Gamma_R = W$ is not of finite index in O(Q). One therefore obtains a class of vanishing lattices, for which the statements of Theorem 5.3.2 and Proposition 5.3.5 are not valid (cf. also Remark 5.3.7).

However, we show that the descriptions of Theorem 5.3.2 and Proposition 5.3.5 still hold for some vanishing lattices (Q,R) corresponding to Coxeter systems of hyperbolic type. Let Q be one of the hyperbolic lattices of Table 5.5.1 (for the notation see Chapter 4.1). Let B be a basis of Q whose Dynkin diagram coincides with the Coxeter graph of Figure 4.1.1 corresponding to Q (with the convention of Chapter 1.5).

Table 5.5.1

Q	Q_{237}	Q_{245}	Q_{334}	Q^2_{2223}	Q^4_{2223}
Q'	E_8	E_7	E_6	D_5	D_7
ε	-1	-1	-1	-1	+1

Proposition 5.5.1. For a vanishing lattice (Q,R) corresponding to a lattice Q of Table 5.5.1, the following statements are true:

(i) $\Gamma_R = W = O^*_\varepsilon(Q)$.

(ii) $R = \{v \in Q \mid \langle v,v\rangle = 2\varepsilon\}$.

Proof. (i) For the first three lattices statement (i) was proven by E. Brieskorn (unpublished). We give his proof, which can be generalized to include the other cases, too.

In each case the lattice Q can be written as $Q = Q' \perp U$ where Q' is a root lattice. The corresponding root lattices Q' are indicated in Table 5.5.1. One can easily show that the lattices Q have the reduction property (R) of [Wall_1, § 5]. Therefore it follows from [Wall_1, 5.2] that

$$O(Q) = O(Q') \cdot O(U) \cdot \Psi_U(Q) .$$

(In [Wall_1] this result is only formulated for unimodular lattices, but the proof goes also through for the lattices Q.) This implies

$$O^*_\varepsilon(Q) = O^*_\varepsilon(Q') \cdot O^*_\varepsilon(U) \cdot \Psi_U(Q) \qquad (*)$$

But $O^*_\varepsilon(Q') = W'$, where W' is the Weyl group of the root lattice Q', and $O^*_\varepsilon(U) = \{1, s_{f_1+\varepsilon f_2}\}$, where f_1, f_2 is a basis of isotropic vectors of U. Here one can take the vectors $f_1 = \tilde{\lambda} + e_{r-1}$ and $f_2 = -\varepsilon(e_r + f_1)$ for f_1 and f_2 respectively, where e_{r-1} and e_r are the vectors corresponding to the two outermost vertices of the longest branch of the Coxeter graph corresponding to Q, and where $\tilde{\lambda}$ is the longest root of Q' [Bourbaki_2]. The group $\Psi_U(Q)$ is generated by transformations

$$\psi_{f_i,\lambda} = s_\lambda \circ s_{\lambda+\varepsilon f_i} ,$$

where λ is a root of Q'. Using these facts, one can easily show that all the groups on the right-hand side of the equation (*) are contained in W.

(ii) Let v be a minimal vector of Q of square length 2ε. We have to show that v is contained in R. Since the lattice Q' has the reduction property (R) of [Wall_1, §5], one can show as in [Wall_1, 5.1] that v is equivalent under $\Psi_U(Q)$ to one of the following vectors: λ, $\lambda + af_1$, $\lambda + bf_2$, or $f_1 + \varepsilon f_2$, where λ is a root of Q' and $a,b \in \mathbb{Z}$. Since $\Psi_U(Q)$ is contained in Γ_R by (i), it suffices to show that these vectors lie in R. By the choice of f_1 and f_2 as in (i) one has:

$f_1 + \varepsilon f_2 = -e_r \in R$. The Weyl group $W' \subset \Gamma_R$ of the root lattice Q' acts transitively on the set of roots. Moreover $s_{f_1+\varepsilon f_2}(\lambda + bf_2) = \lambda - \varepsilon bf_1$. Therefore we only need to show that $\tilde{\lambda} + af_1$ lies in R for $a \in \mathbb{Z}$, where $\tilde{\lambda}$ is the longest root of Q'. Since if δ lies in R then also $-\delta$ lies in R, we may assume that $a > 0$. But now $\tilde{\lambda} + af_1 = (a+1)\tilde{\lambda} + ae_{r-1}$. By applying $s_{\tilde{\lambda}}$ and $s_{e_{r-1}}$ alternately, this vector can be transformed into $\tilde{\lambda}$ or e_{r-1}, which both lie in R. This proves (ii) and completes the proof of Proposition 5.5.1.

BIBLIOGRAPHY

A'Campo, N.:

1) Tresses, monodromie et le groupe symplectique. Comm. Math. Helv. 54, 318-327 (1979).

Arnol'd, V.I.:

1) Remarks on the stationary phase method and Coxeter numbers. Usp. Math. Nauk. 28:5, 17-44 (1973) (Engl. translation in Russ. Math. Surv. 28:5, 19-48 (1973));

2) Critical points of smooth functions and their normal forms. Usp. Math. Nauk. 30:5, 3-65 (1975) (Engl. translation in Russ. Math. Surv. 30:5, 1-75 (1975)).

Beauville, A.:

1) Le groupe de monodromie des familles universelles d'hypersurfaces et d'intersections complètes. In: Complex analysis and algebraic geometry (Ed. H. Grauert). Proc. Conf., Göttingen 1985, Lecture Notes in Math. 1194, Springer-Verlag, Berlin-Heidelberg-New York-Tokyo 1986, pp. 8-18.

Bourbaki, N.:

1) Algèbre Commutative. Chap. 1. Hermann, Paris 1961;

2) Groupes et Algèbres de Lie. Chaps. 4, 5 et 6. Hermann, Paris 1968.

Brieskorn, E.:

1) Die Monodromie der isolierten Singularitäten von Hyperflächen, Manuscripta math. 2, 103-160 (1970);

2) Special singularities - resolution, deformation and monodromy. Mimeographed notes of a series of survey lectures given at the AMS Summer Research Institute Algebraic Geometry, Arcata 1974;

3) Milnor lattices and Dynkin diagrams. Proc. of Symp. in Pure Math. Vol. 40, Part 1,153-165 (1983);

4) Die Milnorgitter der exzeptionellen unimodularen Singularitäten. Bonner Mathematische Schriften Nr. 150, Bonn 1983.

Chmutov, S.V.:

1) Monodromy groups of critical point of functions. Invent. math.67, 123-131 (1982);

2) The monodromy groups of critical points of functions II. Invent. math. 73, 491-510 (1983).

Deligne, P., Katz, N.:

Groupes de monodromie en géométrie algébrique (SGA 7 II). Lecture Notes in Math. No. 340, Springer-Verlag, Berlin-Heidelberg-New York 1973.

Dimca, A.:

1) Monodromy of functions defined on isolated singularities of complete intersections. Compositio Math. 54, 105-119 (1985);

2) Monodromy and Betti numbers of weighted complete intersections. Topology 24, 3, 369-374 (1985).

Ebeling, W.:

1) Quadratische Formen und Monodromiegruppen von Singularitäten. Math. Ann. 255, 463-498 (1981);

2) On the monodromy groups of singularities. Proc. of Symp. in Pure Math. Vol. 40, Part 1, 327-336 (1983);

3) Arithmetic monodromy groups. Math. Ann. 264, 241-255 (1983);

4) Milnor lattices and geometric bases of some special singularities. In: Noeuds, tresses et singularités (C. Weber ed.), L'Enseignement Math. Monographie N° 31, Genève 1983, pp. 129-146; L'Enseignement Math. 29, 263-280 (1983);

5) An arithmetic characterisation of the symmetric monodromy groups of singularities. Invent. math. 77, 85-99 (1984);

Ebeling, W., Wall, C.T.C.:

Kodaira singularities and an extension of Arnold's strange duality. Compositio Math. 56, 3-77 (1985).

Eichler, M.:

1) Quadratische Formen und orthogonale Gruppen (2. Auflage). Springer-Verlag, Berlin - Heidelberg - New York 1974.

Gabrielov, A.M.:

1) Intersection matrices for certain singularities. Funkt. Anal. Jego Prilozh. 7:3, 18-32 (1973) (Engl. translation in Funct. Anal. Appl. 7, 182-193 (1974));

2) Dynkin diagrams of unimodal singularities. Funkt. Anal. Jego Prilozh. 8:3, 1-6 (1974) (Engl. translation in Funct. Anal. Appl. 8, 192-196 (1974));

3) Polar curves and intersection matrices of singularities. Invent. math. 54, 15-22 (1979).

Gantmacher, F.R.:

1) The theory of matrices I, II. Chelsea 1959.

Gibson, C.G., Wirthmüller, K., Du Plessis, A.A., Looijenga, E.J.N.:

Topological stability of smooth mappings. Lecture Notes in Math. No. 552, Springer-Verlag, Berlin - Heidelberg - New York 1976.

Giusti, M.:

1) Classification des singularités isolées d'intersections complètes simples. C.R. Acad. Sc. Paris, Série A, 284, 167-170 (1977).

Hamm, H.:

1) Lokale topologische Eigenschaften komplexer Räume. Math. Ann. 191, 235-252 (1971);

2) Exotische Sphären als Umgebungsränder in speziellen komplexen Räumen. Math. Ann. 197, 44-56 (1972).

Hefez, A., Lazzeri, F.:

The intersection matrix of Brieskorn singularities. Invent. math. 25, 143-157 (1974).

Humphries, S.P.:

1) On weakly distinguished bases and free generating sets of free groups. Quart. J. Math. Oxford (2), 36, 215-219 (1985).

Husein-Zade, S.M.:

1) The monodromy groups of isolated singularities of hypersurfaces. Usp. Math. Nauk. 32:2, 23-65 (1977) (Engl. translation in Russ. Math. Surv. 32:2, 23-69 (1977)).

Janssen, W.A.M.:

1) Skew-symmetric vanishing lattices and their monodromy groups. Math. Ann. 266, 115-133 (1983);

2) Skew-symmetric vanishing lattices and their monodromy groups. II. Math. Ann. 272, 17-22 (1985).

Kneser, M.:

1) Erzeugung ganzzahliger orthogonaler Gruppen durch Spiegelungen. Math. Ann. 255, 453-462 (1981).

Knörrer, H.:

1) Die Singularitäten vom Typ $\tilde{D}$. Math. Ann. 251, 135-150 (1980).

Kronecker, L.:

1) Algebraische Reduktion der Scharen bilinearer Formen. Sitzungsber. Akad. Wiss. Berlin (1890).

Lamotke, K.:

1) The topology of complex projective varieties after S. Lefschetz. Topology 20, 15-51 (1981).

Lê, D.T.:

1) Calcul du nombre de cycles évanouissants d'une hypersurface complexe. Ann. Inst. Fourier, Grenoble, 23(4), 261-270 (1973);

2) Calculation of Milnor number of isolated singularity of complete intersection. Funkt. Anal. Jego Prilozh. 8:2, 45-49 (1974) (Engl. translation in Funct. Anal. Appl. 8, 127-131 (1974));

3) Some remarks on relative monodromy. In: Real and Complex Singularities. Proc. Nordic Summer School Oslo (Ed. P. Holm), Sijthoff and Noordhoff 1977, pp. 397-403;

4) The geometry of the monodromy theorem. In: C.P. Ramanujam, a tribute (Ed. K.G. Ramanathan),Tata Institute Studies in Math. 8, Springer-Verlag, Berlin - Heidelberg - New York 1978, pp. 157-173.

Lefschetz, S.:

1) L'Analysis Situs et la Géométrie Algébrique. Gauthier-Villars, Paris 1924.

Levine, J.:

1) Polynomial invariants of knots of codimension two. Ann. of Math. 84, 537-554 (1966).

Looijenga, E.J.N.:

1) Rational surfaces with an anti-canonical cycle. Annals of Math. 114, 267-322 (1981);

2) The smoothing components of a triangle singularity I. Proc. of Symp. in Pure Math. Vol. 40, Part 2, 173-183 (1983);

3) Isolated singular points on complete intersections. Lond. Math. Soc. Lecture Note Series 77, Cambridge University Press 1984;

4) The smoothing components of a triangle singularity II. Math. Ann. 269, 357-387 (1984).

Looijenga, E., Peters, Ch.:

Torelli theorems for Kähler K3 surfaces. Compositio Math. 42, 145-168 (1981).

Lyndon, R.C., Schupp, P.E.:

Combinatorial Group Theory. Springer-Verlag, Berlin - Heidelberg - New York 1977.

Mather, J.N.:

1) Stability of C^{∞}-mappings. VI. The nice dimensions. In: Liverpool Singularities Symposium I (Ed.C.T.C. Wall), Lecture Notes in Math. 192, Springer-Verlag, Berlin - Heidelberg - New York 1971, pp. 207-253.

Maurer, J.:

1) Puiseux expansion for space curves. Manuscripta math. 32, 91-100 (1980).

Mérindol, J.Y.:

1) Théorème de Torelli affine pour les intersections de deux quadriques. Invent. math. 80, 375-416 (1985).

Merle, M.:

1) In preparation.

Milnor, J.:

1) Singular points of complex hypersurfaces. Ann. of Math. Studies, Princeton 1968.

Nikulin, V.V.:

1) Integral symmetric bilinear forms and some of their applications. Izv. Akad. Nauk. SSSR Ser. Mat. 43, 111-177 (1979) (Engl. translation in Math. USSR Izv. 14, No. 1, 103-167 (1980)).

Pham, F.:

1) Formules de Picard-Lefschetz généralisées et ramification des intégrales. Bull. Soc. math. France 93, 333-367 (1965).

Pickl, A.:

1) Die Homologie der Einhängung eines vollständigen Durchschnitts mit isolierter Singularität. SFB-Preprint, Göttingen 1985.

Pinkham, H.:

1) Groupe de monodromie des singularités unimodulaires exceptionelles. C. R. Acad. Sc. Paris, Série A, 284, 1515-1518 (1977).

Pjateckiĭ-Šapiro, I.I., Šafarevič, I.R.:

A Torelli theorem for algebraic surfaces of type K3. Izv. Akad. Nauk. SSSR, 35, No. 3, 530-572 (1971) (Engl. translation in Math. USSR Izv. 5, No. 3, 547-588 (1971)).

Reid, M.:

1) The complete intersection of two or more quadrics. Thesis, Cambridge University 1972.

Saito, K.:

1) Extended affine root systems I (Coxeter transformations). Publ. Res. Inst. Math. Sci. Kyoto, 21, 75-179 (1985);

2) In preparation.

Siersma, D.:

1) Classification and deformation of singularities. Proefschrift, Amsterdam 1974.

Slodowy, P.:

1) Singularitäten, Kac-Moody-Liealgebren, assoziierte Gruppen und Verallgemeinerungen. Habilitationsschrift, Bonn 1984.

Teissier, B.:

1) The hunting of invariants in the geometry of discriminants. In: Real and Complex Singularities. Proc. Nordic Summer School Oslo (Ed. P. Holm), Sijthoff and Noordhoff 1977, pp. 565-678;

2) Variétés polaires. I. Invariants polaires des singularités d'hypersurfaces. Invent. math. 40, 267-292 (1977).

Wajnryb, B.:

1) On the monodromy group of plane curve singularities. Math. Ann. 246, 141-154 (1980).

Wall, C.T.C.:

1) On the orthogonal groups of unimodular quadratic forms. II. J. Reine Angew. Math. 213, 122-136 (1963);

2) A splitting theorem for maps into $\mathbb{R}^2$. Math.Ann. 259, 443-453 (1982);

3) Classification of unimodal isolated singularities of complete intersections. Proc. of Symp. in Pure Math. Vol. 40, Part 2, 625-640 (1983);

4) Notes on the classification of singularities. Proc. London Math. Soc. (3), 48, 461-513 (1984).

Weierstraß, K.:

1) Zur Theorie der bilinearen und quadratischen Formen. Monatshefte Akad. Wiss. Berlin, 310-338 (1868); Werke II (Mayer und Müller, Berlin, 1895), pp. 19-44.

Wirthmüller, K.:

1) Torus embeddings and deformations of simple singularities of space curves. Preprint, Universität Regensburg 1984;

2) The discriminants of a series of unimodular singularities of space curves. Preprint, Universität Regensburg 1986.

Zariski, O.:

1) Algebraic surfaces. Second supplemented edition. Ergebnisse der Mathematik, Bd. 61. Springer-Verlag, Berlin - Heidelberg - New York 1971.

SUBJECT INDEX